A UFO WENT DOWN.
THE EVIDENCE IS IN.
AND THE MYSTERY CONTINUES.

❖ A glittering yellow foam covered the crash site—seen,
collected, and reported by numerous fishing boats that
had sailed to the "rescue." What was that strange
foam?

❖ Debris was brought up from the harbor floor—an
event witnessed by onlookers, performed by navy
divers, and verified by Ledger and Styles. Where is
that debris?

❖ In nearby Shelburne, site of a top-secret submarine
base, the military blocked the roads, and a flotilla
of naval ships steamed into the area's waters.
Underwater photos were taken. What did those pictures reveal?

And why is the incident still being hushed up?
FIND OUT IN . . .

DARK OBJECT

Other books by Don Ledger

The Maritime UFO Files
Swissair Down

DARK OBJECT

THE WORLD'S ONLY GOVERNMENT-DOCUMENTED UFO CRASH

BY **DON LEDGER** AND **CHRIS STYLES**

INTRODUCTION BY **WHITLEY STRIEBER**

A DELL BOOK

Published by
Dell Publishing
a division of
Random House, Inc.
1540 Broadway
New York, New York 10036

Dell books may be purchased for business or promotional use
or for special sales. For information please write to:
Special Markets Department, Random House, Inc.,
1540 Broadway, New York, N.Y. 10036.

Dell® is a registered trademark of Random House, Inc., and the colophon
is a trademark of Random House, Inc.

ISBN: 0-440-23647-9

Manufactured in the United States of America

Published simultaneously in Canada

April 2001

10 9 8 7 6 5 4 3 2 1
OPM

The authors would like to dedicate this work to all UFO witnesses, in recognition of their courage in coming forward and sharing their observations and convictions, sometimes at great personal risk.

The authors can be reached at www.donledger.com.

INTRODUCTION

by Whitley Strieber

The Shag Harbor incident is something totally unexpected: a UFO case so flawlessly documented that it *must* have happened. But it is hardly known, even among UFO investigators. Perhaps it has been discreetly covered up. Perhaps it has fallen through the cracks.

Either way, Don Ledger and Chris Styles have spent years working with this profoundly convincing and important story, and in this book they offer the most complete and authoritative account of the event that has ever been attempted.

Did a spacecraft from another world plunge into

the cold northern waters of Shag Harbor on an October night in 1967?

The authors offer the most convincing evidence that has ever been presented for a UFO crash. They actually prove that the event took place, and they do it using government documents.

The documents in the appendix of this book are only four of the many official reports that were discovered by Styles and Ledger. They reveal that the sighting of a UFO by local fisherman Leo Mersey was actually taken so seriously by the Royal Canadian Mounted Police that they filed an official report. They took the sighting report of two brothers, Lawrence Charles Smith and Norman Eugene Smith, equally seriously, as is evident from the report they filed based on their testimony.

There were reports of mysterious lights all over the Shag Harbor area, and extensive official interviews of witnesses. Indeed, concern was so strong at high levels within the Canadian government that a Royal Canadian Air Force discussion of the incident suggests an underwater investigation of the site.

The crash took place, without a doubt, and the government took it very seriously indeed, as the documents in the appendix demonstrate. Indeed, if this had been a crash of anything except a UFO, the fact would have been duly noted and a search would have been initiated for survivors. Instead, a period of confusion followed, and secrecy so intense that even years later Ledger and Styles were tailed by police after they took pictures of the area where the incident took place.

Why would this be? If an unknown craft definitely crashed—and this seems impossible to refute—then why was the official reaction so strange? To understand this it is necessary to understand first that the human response to the UFO phenomenon is extremely strange and distorted.

For example, the documents referred to in this book prove beyond a shadow of a doubt that the Canadian government knew for certain that something unknown crashed into Shag Harbor. But the official response to the crash, of first doing nothing and then refusing to confirm anything, is wildly inappropriate.

That the object was not any sort of secret aircraft is made clear by the witness testimony of Captain Pierre Charbonneau, pilot of Air Canada Flight 305. He appears to have been the first witness to the object, which he saw from his plane. He described the object as orange and rectangular in shape, followed by a string of smaller lights. The object appeared to be huge.

An astonishing incident then occurred: there was an explosion near the large object. Another explosion took place a few moments later, then the trail of lights began some strange maneuvers.

This sighting alone would have entered the annals of UFO history, as it appeared to show some kind of accident or battle among the strange objects. But that was only the beginning of this incredible incident.

Since it led, a short time later, to a crash that was witnessed by multiple parties and even openly admitted by the government to have involved a UFO, one would think that there would have been headlines worldwide. Aliens hadn't exactly landed, but they had certainly crashed.

Instead, the incident was quietly filed away.

On the surface what happened at Shag Harbor in October of 1967 is straightforward enough: a large object that had been seen, both by ground observers and pilots, crossing the sky in apparent distress went crashing into the water in full view of a substantial number of other witnesses. Strange foam and odd debris were sighted. The incident was reported to the police, the military investigated, and government documents openly acknowledged that the incident involved a UFO. Divers brought up material, again in full view of witnesses.

What is not so straightforward is that the government never released any information about what was found beneath the dark waters of Shag Harbor.

The government response became extremely strange and secretive, and very different from what it would have been if this had been an aircraft. There was no extended search-and-rescue operation, not in any ordinary sense. There was never any explanation from the government about what had happened.

The only reason that the Shag Harbor incident came to light is that Chris Styles and Don Ledger would not let the matter rest. And why should they? This is the best documented UFO crash incident that is publicly known. After October of 1967 it disappeared from memory, not returning until Ledger and Styles began their investigation and wrote this book.

A question begs to be asked: why, given that the government itself said that an event of phenomenal importance had occurred, was it simply ignored afterward and held secret? After all, this book offers documentation that the incredible has happened: an unidentified flying object of unknown origin crashed. Not only that, debris was recovered. If the object had been a meteor, there would have been no reason to conceal that fact. Witness descriptions confirm that it was not an aircraft in any ordinary sense of the word.

So what happened here? Somebody must know something, because somebody, before numerous witnesses, did indeed recover materials from the floor of the harbor. But the public hasn't been told a word.

This is not the only UFO event that has been cloaked in extreme secrecy. It is simply the only one in which there has been a completely irrefutable admission by government authorities that a UFO was involved.

Why would the Canadian government, or the U.S. government, or any government, for that matter, be so extremely secretive about a UFO incident? In fact, why is there any secrecy about UFOs at all? The Shag Harbor incident not only proves that UFOs exist and that they can crash or at least plunge violently into the sea, it also proves that the government knows this and does nothing about it—at least, nothing that is revealed to the public. Understand, the incident *proves* this. Shag Harbor is no ordinary UFO case. It is the best documented and one of the best witnessed of all UFO cases.

It means that UFOs are almost certainly real. It appears also to mean, based on the responses of the Canadian government, that governments must in general be extremely secretive about them. Certainly only a few governments have ever taken any official position that UFOs are, if not real, then worth studying. These governments are Chile, Ecuador, China, and France. But none of them have ever released any definitive proof.

Dark Object certainly offers definitive proof that something very, very strange plunged into Shag Harbor in October of 1967. There is additional evidence, covered in riveting detail in this book, that it maneuvered underwater for a time.

It is a given that government secrecy exists in the UFO area. But there is a larger question here. Governments notwithstanding, why are the alleged aliens themselves so secretive? No alien has ever landed in public. No alien has ever communicated with us openly. If they are here, they are obviously able to do so if they choose.

In fact, if they *are* here, then it would seem that government policy and alien policy are in agreement: their presence must be kept a deep, dark secret.

There are two reasons that aliens might have a policy of secrecy. The first is that they might be doing something that we might resist if we understood it. Under this scenario they would be coercing the government into silence.

However, there might be another reason entirely. In a paper entitled "Searching for Extraterrestrial Civilizations" by T.B.H. Kuiper and M. Morris, published in *Science* (Vol.196, May 6, 1977), the idea is put forth that an advanced extraterrestrial civilization would almost certainly be extremely secretive if they came here. The authors' reason for this is cogent. "Before a certain threshold is reached, complete contact with a superior civilization would abort further development through a culture shock effect." The authors go on to say that an advanced civilization would probably have nothing to gain from us except fresh new ideas, and that they would have to remain hidden until we had undergone a gradual process of acclimatization or risk the destruction of independent human thought.

The same thing would happen to us that happened to dozens of tribes worldwide beginning in the sixteenth century, as European technological civilization advanced. Even those that were not forcibly destroyed, such as the Aztecs and the American Indians, experienced the complete redirection of all of their intellectual, spiritual, and economic activity toward the process of obtaining western goods.

This is why, by the mid-twentieth century, anthropologists were making massive efforts to keep newly discovered tribes isolated so that they could be studied without the destruction of their cultures.

Something like this could be going on here. It would certainly explain the reticence of the aliens, and the ferocious effort that government makes to keep their secrets, knowing that the penalty for revealing their presence too soon could easily be the abandonment of their interest in mankind.

If, for whatever reason, aliens have forced the U.S. government to help them conceal themselves, it might explain what is, after all, an extremely unusual policy on its part. There is more than adequate evidence that UFOs are a genuine unknown. There is no rational basis for official secrecy, media debunking, and the rabid denial that forms a basic tenet of the modern culture of science. Any one of a hundred UFO cases could form the basis for serious investigation, and the Shag Harbor incident, with its detailed evidence of the operation, and possible failure, of technology, would be an excellent place to start.

In the early years it appeared that UFO policy might go either way. Captain Edward Ruppelt became chief of Project Grudge, which soon became Project Blue Book. Ruppelt, with a degree in aeronautical engineering, brought a high level of technical expertise to UFO investigation, and Blue Book gathered and documented many extraordinary cases from 1951 until 1959. Over this eight-year period the purpose of Blue Book gradually changed. Ruppelt, with his eager desire to solve the problem and inform the public, was replaced by one manager after another, each less inclined than the last to follow Ruppelt's original design.

In 1963 the project was taken over by Major Hector Quintanilla. By this time Blue Book was nothing more than a public relations effort. In 1989, when I spoke to Major Quintanilla about his work on the project, he said that he had done nothing but gather data sent in by the public, and had no personal opinion about UFOs. He was very different from dynamic and effective officers like Ruppelt, who had made serious efforts to at least gain scientifically sound information rather than passively file reports without any real investigation.

In 1965 a massive UFO wave occurred worldwide, and it was during the three-year aftermath of this wave that the Shag Harbor incident took place on the night of October 4, 1967.

It was during this period that the air force began aggressively downplaying sightings. During a previous wave in 1952 the air force had received substantial public criticism for its inability to assert U.S. control over its own airspace, which is its most fundamental mission.

The 1965 wave brought a hardening of the air force position. Ruppelt and his approach were forgotten. The official word was that UFOs were a mystery not worth exploring.

In 1969 a disaster took place for UFO research, one from which it has never recovered. A committee of distinguished scientists had been asked by the air force to evaluate Project Blue Book. Their final report recommended that the air force drop UFO investigations. The highly respected name of Dr. Edward U. Condon carried a lot of weight, and the scientific community accepted the conclusions without ever noting that the report upon which they were based contained a substantial number of unknown or unverified sightings.

The report had vocal critics, including no less an authority on official secrets than Admiral Roscoe Hillenkoetter, the first director of the CIA. He was adamant that UFOs deserved serious investigation. In

1960 he had stated, "Unknown objects are operating under intelligent control."

In 1968 the Congressional Committee on Science and Astronautics held a symposium on UFOs at the request of Representative J. Edward Roush of Indiana. Representative Roush indicated that it was his belief that the Condon Committee was being improperly influenced by the air force, and that its report was a foregone conclusion.

Six distinguished scientists attended the symposium, including sociologist Dr. Robert L. Hall, prominent UFO investigator Dr. J. Allan Hynek, Dr. Carl Sagan, and Dr. James McDonald, who would eventually commit suicide, apparently over his treatment by the scientific community. Two engineers, Dr. James A. Harder and Dr. Robert M. Baker, also sat on the panel.

The symposium members stated that UFOs deserved serious study. Dr. Harder stated, "On the basis of the data and ordinary rules of evidence, as would be applied in civil or criminal courts, the physical reality of UFOs has been proved beyond a reasonable doubt."

But the conclusions of the symposium were ignored in favor of publishing the Blue Book report, by which the U.S. government officially abandoned even the collection of UFO data, let alone any research into its meaning.

From 1969 until now no case has emerged or incident transpired that might force the government to admit that, on some level, UFOs represent a genuine mystery.

The Shag Harbor incident has the potential to be

such a case. The official record of this crash was made public. In addition, the whole incident was observed in detail by many witnesses. And it did not involve something as simple as a glimpse of an object plunging into the water. The approach of the object was slow and involved maneuvers that were carefully observed by trained professionals.

Indeed, as the web of witness and documentation is woven over the course of this book, it will become apparent that there is, quite simply, no better UFO case in the world. If this were the only UFO incident that had ever happened, it would represent all but certain proof that something unknown was involved.

There is, in addition, a most unusual twist to this story. It involves one of the authors of this book.

Above the dark waters of Nova Scotia on a cold October night, an orange light is observed by many different people. One of those observers leaves his home and goes racing down the street to get a better look. Transfixed, his blood racing, he ends up having a breathtaking encounter with the object at close quarters. Perhaps that is why this book is so immediate and intense.

This particular eyewitness to what became the Shag Harbor incident was *Dark Object* coauthor Chris Styles. So this is not only a carefully researched examination of one of the world's most compelling UFO incidents, it is also a firsthand account by a man who was there.

There is probably only one aspect of the UFO phenomenon that the entire scientific community can agree on. If extraterrestrial craft truly exist in a physical sense, they are, without doubt, the most exciting event in recorded history. Scientific treatments of UFO phenomena often take a dim view of the possibility that anything extraterrestrial is responsible for unexplained sightings of UFOs. The public is less sure.

A 1987 Gallup poll indicated that fifty percent of Americans are convinced that UFOs exist. Ten percent, or 26 million, claim to have had sightings or encounters. In Canada the percentages are slightly higher. Yet many scientists still do not acknowledge that the phenomenon exists. Ironically, some of those same skeptics listen with powerful radio telescopes for whispers from other star systems in a universe they believe must be teeming with life.

This book tells the story of the crash of a highly unusual object into the waters of Nova Scotia's southwest shore on the night of October 4, 1967. No aircraft were reported missing. No space junk was tracked reentering our atmosphere, and natural phenomena were ruled out due to reliable observations from the many professional and credible witnesses. The case has remained unsolved for decades, despite an extensive international search effort by the military.

The crash of the sixty-foot "Dark Object," as it is referred to in many government documents, was not taken lightly by the government. Documents discovered during the investigation of this mystery revealed Canada's hopes of retrieving and accessing alien technology. In the 1960s Ottawa had great expectations that major discoveries might be gained from UFO phenomena. Some people have told me that the Department of National Defense suffered from "Uforia."

Dark Object is also a book about the people, in and out of uniform, whose lives were affected by an encounter with a UFO on that clear dark night in October 1967. Many had to adjust their personal belief systems to include events for which they had not been prepared. Decades later they would still like some answers.

Chris Styles
Halifax, Nova Scotia

I first became involved with this project after meeting Chris Styles. His enthusiasm for this incident was and is contagious.

For many years I have been interested in the UFO phenomenon, reading just about anything I can lay my hands on. I have always been prepared to give the benefit of the doubt to those who are willing to go on the record with their sightings or encounters. But as I grow older, I've found myself growing more skeptical (Chris has already accused me of having the heart of a debunker), not necessarily of the witnesses but of what some writers are prepared to accept as solid evidence. Indeed, I am a little surprised at how easily some investigators embrace a theory or statement that helps to reinforce their own beliefs, regardless of how little evidence there is to sustain it.

I took the step about three years ago of becoming involved in the investigation of these phenomena and found it to be fascinating. The people I have had the opportunity to interview are serious individuals who, I am certain, truly believe they have had a close encounter of the first, second, third, or fourth kind. I have no reason to doubt them. But each incident has been limited by the available evidence or the number of observers on hand.

This book does not deal with one or two individuals but dozens, if not hundreds, who witnessed the incident. And what makes it fascinating is that the closer you get to the subject matter, the more evidence you uncover. This does not hold true in some of the more celebrated UFO cases, where the evidence and witnesses seem to melt away when serious investigative practices are applied.

Chris Styles has done an admirable job of digging out the evidence, separating the speculative from the hard testimony, and tracking down witnesses long gone from the area. This kind of research runs up telephone and travel expenses at an alarming rate, as I can testify to with my own credit card receipts, and there is no hope of compensation except for a sense of personal satisfaction. Chris has managed to accomplish a great deal of research, despite his own limited financial resources, as this book will show.

Now I find the same thing happening to me. The more I dig into this thing the more I discover. When I uncover witnesses they put me in touch with others who were involved or affected in one way or another.

This incident is probably the best documented of all of the UFO crashes to date and that is its strength. The government documents and private testimony that have been gathered can leave no doubt in the reader's mind that this event was real.

And there is a dark side to it, too, with chilling overtones. This may be why the military referred to this artifact as the "Dark Object."

Don Ledger
Bedford, Nova Scotia

THE BEGINNING

DON LEDGER

Every journey has a beginning. This book was begun during a 1995 trip with Chris Styles in his Chevy Blazer.

I met Chris after I joined the UFO organization MUFON earlier that year, after reading countless books about the phenomenon. MUFON put me in touch with Chris, who was then the assistant provincial director in Nova Scotia.

Chris brought up the subject of Shag Harbor during our first meeting. His enthusiasm for the case was infectious, and over the next few months I found myself being drawn into the investigation.

On this sunny day in 1995 I had accompanied

Chris to a meeting with the president of Canadian Seabed Research in Porter's Lake, Nova Scotia. I went along as an interested party, and as a possible member of an underwater survey team in Shag Harbor, Nova Scotia, the next summer.

There was also another reason, one of convenience. In the weeks prior to this trip, as a fledgling Mutual UFO Network (MUFON) investigator, I had been looking into a UFO sighting by two fourteen-year-old girls one year earlier, on the thirteenth of April, 1994. At the time the sighting had received some attention by the local press and electronic media. One of the girls lived in a suburb of Dartmouth known as Cole Harbor, a community in the same general area as Porter's Lake.

During the initial investigation I didn't have my camera with me, and I wanted to photograph the area and get an idea of the location of the sighting. Since the area in question would be just a little out of our way, I asked Chris the night before if he would mind going there on our return trip, and of course he agreed.

After the meeting, we arrived at the sighting location around three-thirty in the afternoon. The area was new to Chris. While I took some pictures, he looked around. It was while I had the camera to my eye that he noticed that I was not the only person photographing this spot. There was another car, a black unmarked Nissan, parked farther down the road in this sparsely populated location. Chris observed a man with a camera shooting in our direction, either at us or at the wooded area I was photographing. I glanced toward the vehicle too late to see the other photographer. "Probably a real estate agent," I suggested. "There's a lot of land around here for sale." I continued snapping pictures.

The car started and proceeded up the road in our direction. By this time I was busy nailing down a compass bearing. The Nissan passed us and turned left around a corner. The road ended there, and there was only cleared land, with forest after that.

"Hey! GRC!" Chris exclaimed. He was referring to the Gendarmerie Royale du Canada, otherwise known as the Royal Canadian Mounted Police, or "Mounties."

"What?" I asked, looking up from my compass.

"Those guys had GRC shoulder patches on their jackets. I saw them when they went by us."

"Are you sure?" I asked. I watched as the vehicle continued down the road to our left, eventually disappearing behind some trees in the thickly wooded area.

"Saw them as plain as day," Chris said. "They weren't more than ten feet away when they went by."

"Why would they be photographing the same area as me?" I was puzzled. "I'm sure they would have done all of that stuff last year, when the UFO was reported, if they were even interested then."

"Could be they were taking pictures of us."

"Us . . . why us?" I was frankly shocked by his suggestion.

"Beats the hell out of me. But I'm less and less surprised by the things that happen in this field, as I get further into it."

"Bull!" I snorted. I wasn't ready to buy into that yet.

We got back into the truck and began the drive back to the city. Chris continued talking, relating some of his own experiences of odd coincidences that had occurred since he had become involved with the incident of the Dark Object. It was then that he said he was considering writing a book about his adventures, if he could find someone to help him. I mentioned that I did some writing, mostly on aviation-related subjects. We agreed that we would talk about the idea over coffee later in the week, and continued the drive back to the city.

My thoughts went back to the road in Cole Harbor and the GRC shoulder patches. I wondered then if they had taken my picture and if so, why.

CHRIS STYLES

One evening when I was twelve years old, I sat in my bedroom listening to the radio and practicing my guitar, thinking perhaps someday I would be the lead guitar in a famous rock band. Soon the rock music ended and I switched off the radio. It was 10:00 P.M., and I could hear my parents moving around downstairs.

My house was on a corner, halfway up a hill in Dartmouth, Nova Scotia. My room was upstairs, in the back of the house, with a view of one of the world's best natural harbors, Halifax Harbor. Bored, I looked out my window, down toward the mouth of the harbor. It was dark now, early in the fall of the year. Lights ringed the harbor and drifted along the water on ships and boats.

One light grabbed my immediate attention. It was a round object, glowing orange, the color of iron heated in a forge. It was much bigger than the others, and was drifting up the harbor toward me. I knew at once that this was no ordinary buoy or running light. There was something both familiar and foreign about it.

Although reluctant to lose sight of it, I wanted to get a better look, so I dashed into an unused bedroom, grabbed my grandfather's old field glasses from a shelf in the closet, and ran back to my room.

My hands shook as I tried to focus the binoculars on the orange disc. I could see no form behind the orange hue. It grew larger as it drew closer but I realized I was going to lose sight of it behind the buildings along the water's edge.

An urgency gripped me. I had to see more.

In desperation I threw open the door to my room and ran recklessly down the stairs. When I reached the front door, I yelled to my parents that I had to get something at the store. I heard my father call after me, telling me to get back in the house in a commanding tone of voice. At any other time this would have stopped me short, but not now. An overpowering curiosity had taken hold of me, and there was no time to explain. I ran through the door, rounded the corner, and was soon running down Pinehill Street as fast as I could.

I reached Prince Albert Road, turned right past St. James Church, and darted down the shallow grade of Canal Street. Over the top of the Mayflower Club I could see the object drifting closer to my end of the harbor.

A whirl of emotions coursed through me. First among them was fascination mixed with fear. Something inside, my own common sense, was telling me to turn back. But the sense of wonder overcame all that. I ran through a vacant lot, my sneakers crunching on gravel, taking a shortcut over to the next street.

Breathlessly I came up from behind the last of a row of warehouses and onto a deserted street that terminated at the railroad tracks. There I stopped short, out of breath, sucking in large gulps of air.

It was there, drifting toward me noiselessly from my right, just above the water's surface. It was an opaque featureless ball that glowed a dull orange. Up until then I hadn't realized how big it was—it was easily fifty or sixty feet in diameter.

I stood rooted to the spot, fascinated, though now apprehensive. Twelve-year-old bravado was being replaced with twelve-year-old imagination. Suddenly I wondered, was this coming for me? Did it know I was here? Should I run and risk attracting its attention, or stay put, hoping it would pass me by? Or did its occupants even care about the presence of a young boy?

Closer inshore it came, tracing the shoreline, almost abreast of me, seventy-five or a hundred feet away. I stayed put, watching it, wondering if it was watching me. Were there beings in there who could see me, or was it some kind of probe?

Now it was in front of me. I looked around. Was anyone else seeing this? After all, this was not an unpopulated area. Thousands of people lived nearby in the houses and apartment buildings that overlooked the harbor. To my left, three or four hundred yards away, were the Department of Transport offices and the docks, with coast guard vessels tied alongside.

Excitement and the cool fall air made me shiver. I was overwhelmed by what I was seeing. All of my instincts told me it was not natural, not fashioned by human hands, not humanly controlled. It drifted across in front of me to my left, turning with the bend of the cove and gliding silently along toward the coast guard complex.

I had been glued to the spot, but now my nerve broke. I turned and ran home as fast as I could, my heart pounding with fear. I kept looking back to make sure that this thing was not following me. I was elated and perhaps a little disappointed that it didn't follow me home.

Finally I reached the house. Now I had to deal with my father's wrath. My story fell on deaf ears. He either did not believe or was unable to accept what his son had witnessed.

I went to my room and again looked out of the window and down the harbor, but the object was no longer there. I sat on the edge of my bed, wondering. The next day the radio reported many calls from around the harbor area, telling about the strange glowing orange ball out on the water the night before. Back then I had no way of knowing how this sighting was going to affect my future and how it would draw me into the world of UFO phenomena. Nor did I suspect that the fascinating events that were unfolding that evening of October 4, 1967, would culminate in an incident, in a little fishing village on the south coast of Nova Scotia, that would profoundly affect me and the lives of so many others.

CHAPTER TWO

THE NIGHT OF THE UFOs

October 4, 1967, 7:19 P.M. Atlantic daylight time
Douglas DC-8, Air Canada Flight 305
On Victor 300, level at 12,000 feet
Between Sherbrooke and St. Jean VOR
Southeastern Quebec

Captain Pierre Guy Charbonneau relaxed in his seat in the cockpit, his eyes scanning the instruments on the panel before him. He took comfort from them, their dials dimly lit by red lights to protect his night vision. Each was reading exactly what it should be reading. He was on an IFR (instrument flight rules) flight plan and could have flown the entire trip keeping his eyes inside the cabin, trusting in these very instruments.

The slipstream whispered by the windscreen of the DC-8 at 256 knots. Everything seemed in order. The first officer, Robert Ralph, was beside him.

He allowed his gaze to wander outside the cockpit to the crisp, clear, and starry night sky. It was a good night for flying, with only a thin wispy cloud layer just below their altitude of twelve thousand feet. He glanced out of the window near his left shoulder, looking to the south.

What he saw there made him sit up straighter in his seat. A well-lit, large, orange rectangular object, followed by a string of smaller lights, like the tail on a kite, was tracking parallel to their course at their altitude about twenty degrees above the horizon. The brightly lit object appeared to be huge.

Captain Charbonneau immediately drew it to the attention of his first officer, Ralph. It was standard practice to alert each other about any other traffic in their vicinity. But this was something different, and they speculated about what it might be. Neither had ever seen anything like it before.

Suddenly, at 1919 hours ADT, there was a sizable explosion near the larger object. They watched in amazement while this explosion turned into a big white ball-shaped cloud that quickly turned red in color, then violet, and then blue. The captain planted his feet on the rudder pedals and gripped the control yoke, ready to override the autopilot and take control if necessary.

Two minutes later, at 1921 hours, there was another apparent explosion that turned into a second sphere, which was orange in color. This was bigger and higher up than the other but like the first, it, too, eventually faded to blue. The smaller lights on the "kite tail" broke formation with the rectangular object and began to dance around the spheres like fireflies.

The captain let out his breath, realizing he had been holding it for some time. His first instinct was to insure control of the aircraft, checking airspeed, altitude, and course, but the autopilot was doing its job properly.

The captain and first officer kept an eye on the light show in the air for several more minutes, watching as the second pear-shaped cloud, glowing pale blue, drifted eastward. "What did you make of that?" Ralph asked. They had been speculating as to what the object was during the sighting.

The captain shook his head with a puzzled expression. "I can honestly say I've never seen anything like that before."

"Do you suppose it could have been some type of military aircraft?" Ralph asked. "Maybe it was something they were testing out of Loring Air Force Base in Maine. That's not too far southeast of here."

Again Captain Charbonneau had to plead ignorance. "Your guess is as good as mine, son."

The first officer paused for a moment, then asked the question that hung in the air between them. Somebody had to bring it up.

"Should we file a report on this?"

The captain glanced over his left shoulder to the

now empty sky to the south. *Weird,* he thought. He
pursed his lips, turning his attention back to his instruments while he mulled over that question. It was an
important one for both of them. For a moment Ralph
thought he hadn't been heard, but finally the captain
answered.

"Yeah, I think we should, don't you—something
that close?"

Ralph shrugged. "You're the boss. What should
I say?"

"Just tell them what you saw."

The first officer thought this over. What had he actually seen? Like most professional pilots he was reluctant to discuss, admit to, or report a UFO for fear of
being ridiculed or, worse, having it affect his career.
And there was another reason, a regulation being en-
forced by the Canadian government of which few
civilians were aware. It was a regulation that stemmed
from an agreement between Canada and the United
States that restricted the divulging of UFO reports beyond the guidelines set forth in and stemming from the
JANAP 146 (d) agreement of 1962.

Although the guidelines were extensive, essentially they restricted the reporting of UFOs by military
and airline pilots, requiring them to report their sightings to the authorities only. They were not to report
them to civilian agencies, the press, or anyone else, under penalties that could result in a fine of ten thousand
dollars and a prison term of up to ten years. Ralph
voiced this concern.

"My advice is just to write down what you saw. Give

only the facts. I'll do the same. We can request anonymity from the company and have them pass it on."

That seemed to cheer up the first officer.

"Yeah, maybe that would be the best way to go."

Despite his concerns First Officer Ralph wrote up a fairly comprehensive report (as did his captain) with noted times and drawings. He did this despite the fact that at the time he was in training to become a captain with Air Canada.

7:51 P.M. ADT
Eastern Passage, Nova Scotia
Outskirts of Canadian Naval Air Station, Shearwater

William Thibeault and his brother were set up for stargazing on William's front lawn using a surveyor's transit as a telescope. Both brothers were familiar with the various conditions of the sky. William was an employee of Fairey of Canada Limited, a subsidiary of Fairey Aircraft in England, the company responsible for the design of many modern aircraft.

It was a cool but clear evening with no moon, nearly perfect conditions for viewing the sky. Just across the harbor and about one and a half miles to the southwest of them was Halifax, the capital city of Nova Scotia, with its waterfront and office buildings all lit up, but there was not enough backscatter to interfere with their sky watching. They noted a layer of cloud at twelve thousand feet and soon noticed something else even higher. They saw two dim white lights followed by a brighter white light, all moving slowly in a northeast to southwest direction. This course would take them down the coastline of Nova Scotia, over towns and villages, in the direction of Shag Harbor.

The two brothers watched the mysterious lights for about five minutes, as they transited the nearly clear night sky, and speculated about what they might be. They excitedly reported them to the Naval Air Station at seven minutes after eight. When asked by the commanding officer at the base in Shearwater for an estimate of their height, William gave an estimate of fifty to one hundred thousand feet.

When asked what they appeared to be, he replied that he didn't know, but he was certain that they were not aircraft.

9:00 P.M. ADT
Aboard the MV Nickerson
32 Nautical Miles south of Sambro, Nova Scotia

Captain Leo Mersey's patience was wearing thin. He turned to the first mate. "Let's get these men below." He didn't like it when the men behaved like kids. "This is a fishing dragger, not a cruise liner."

With the crew, numbering eighteen, crowding the ship's rails, there was danger of one of them falling overboard, especially when they were excited like this.

"I want someone to answer when I call below."

"Aye, Captain. Come on, boys, you heard the cap-

tain." The first mate followed the men back to their
posts and their bunks. When he was assured that all
was in order, he made his way back to the wheelhouse.

Captain Mersey was looking at the screen on the
Decca radar. The mate and the captain had a good
working relationship and he could be less formal when
they were away from the men.

"Are they still painting targets on the scope, Leo?"

Mersey turned to him. "Yeah. All four of them.
Looks like they're holding their position about sixteen
miles northeast of ours."

"They're real, then. What the hell do you think
they are?"

The captain looked off to the northeast, where they
could see four brilliant red lights that appeared to be
on or just above the water. They were spaced out in a
box formation, about six miles on a side.

"They're real and solid enough if the radar is
painting them. As to what they are, it beats me. The
navy does exercises in that area, but I never saw anything like this before. In fact I can say I never saw anything that red before."

Occasionally one of the red lights would flare up
to such an intensity that it would leave an afterimage
when they averted their gaze. It was an astonishing effect. "Maybe it's some kind of flare the navy is trying
out, Leo. They don't always give us warnings."

Mersey shook his head. "No. When you were on
the foredeck with the crew, I radioed into RCC and the
harbormaster in Halifax. They don't have exercises on

tonight. Besides, the radar paints them as being pretty substantial. It wouldn't show flares and they don't stay up there like that."

He squinted his eyes as one of the lights flared up again. "There's something there, all right, and I'm content to keep my distance."

They were interrupted when one of the younger hands knocked on the door of the wheelhouse and entered. "Sir, I thought you might like to know something I just heard on the multiband radio."

Mersey cocked his head. "What is it, son?"

"Well, things sound pretty crazy on the mainland. I guess the Mounties are getting complaints about UFOs from all over. It's really thick from Halifax to Yarmouth."

Mersey nodded and looked again at the lights. He turned back to the crew member. "Thanks, son. Let me know if you hear anything solid." He considered telling the youth to keep it to himself, but thought better of it. Despite the workload on a dragger, there was always time to talk, and telling him that would only encourage him.

"Leo, have a look at this!" Mersey's attention was immediately drawn back to the lights by the urgency in his first mate's voice. It was not like him to show excitement so plainly.

One of the red lights was climbing upward, arcing toward them. It passed directly over the ship at what Leo guessed was over a mile up and continued toward the horizon.

The pair stared in silence for a moment, the only

sounds coming from the creaks and groans of the dragger's gentle rolling in a long sea swell. This was soon shattered by the ship's radio.

"MV *Nickerson,* MV *Nickerson,* MV *Nickerson,* this is Coast Guard Radio Halifax, Coast Guard Radio Halifax, over."

Mersey started in spite of himself. He grabbed the microphone, thumbing the switch.

"Coast Guard Radio Halifax, this is MV *Nickerson.* Go ahead."

"MV *Nickerson,* I have a message for Captain Leo Mersey. Over."

"This is Captain Mersey, over."

"Captain, I have a message from RCMP Headquarters in Halifax. It requests that you report to their detachment in Lunenburg when you finish your trip. They would like a report on your sighting. Over."

The captain affirmed that he would do so and ended his transmission.

"It seems, Leo," the first mate said, "that someone was listening in on your transmissions to RCC and the queen's harbormaster in Halifax."

The captain muttered his agreement. He seemed lost in thought.

Water Street, Halifax
10:00 P.M. ADT

The glowing orange-colored fireball caught the woman's attention from the Halifax side of the harbor in the capital city. She watched from her position near the ferry wharf, fascinated by the orb's effortless passage from the east side of the harbor to the boat slips. Minutes later, at home, she watched it for several minutes through a pair of binoculars. The object drifted into Dartmouth Cove, curved back toward the coast guard pier, then went southeast toward the Imperial Oil refinery. At that point she lost track of the sphere.

The woman called a local radio station and reported her sighting in an excited voice. She described it as an object about forty to fifty feet in diameter, glowing orange, the color of a red-hot poker. She didn't leave her name.

Not long after, the station received another call confirming the first caller's description. They would get many such calls that night, as did the local newspaper, *The Halifax Chronicle Herald.*

While the woman was watching the object, a mile away, across the water, young Chris Styles watched the same object, unaware that he was not the only one who saw it.

Somewhere between 10:30 and 11:00 P.M. ADT
Mason's Beach in Puffy Cup Cove
Lunenburg Village, Nova Scotia

Will C. Eisnor is a professional photographer and owner of Knickle's Studio in Lunenburg, a lovely little fishing port. On the evening of October 4, 1967, one of his schooners was about to have her life ended forever.

His forty-year-old, thirty-foot sailboat was in such bad shape that she was beyond repair. He had pulled her close to shore and as high up on Mason's Beach during high tide as possible. Will and two of his friends, Raymond Hiltz and St. Clair Croft, spent some time placing fuel aboard the vessel so that they could destroy her by fire.

They had chosen a good night for a big fire, a cool evening with no wind. There was no moon, but the sky was filled with stars. Little wavelets lapped up on the shore while the tide first ebbed, then began to flow back in.

Burning the craft was not to be as easy as they had anticipated; inside and out her planks and timbers were sweetened by years of being submerged in the waters of the Atlantic Ocean. They made many trips up and down the beach, bringing back armloads of driftwood to add to the fire, trying to dry the vessel's structure enough for it to burn on its own.

It was during one of these trips down the beach that Will noticed some lights in the night sky to the west. They stood out against the star field and were very arresting. He stared at them for a while, trying to figure out what they might be. They were too bright and too colorful to be stars, and they appeared to be suspended, motionless, over a ridge covered with spruce trees, although the photographer suspected that they were farther away than that.

There were three lights, arranged in a tilted triangle. There was an amber light at the bottom right of the triangle, a brilliant blue light at the apex, and what looked like a spear of amber light to the rear on the remaining corner. For the life of him he could not imagine what it might be. There was no sound coming from it. Instinctively he knew this had to be some strange, unknown phenomenon. He wondered how long it had been there.

He returned to the boat and mentioned it to his friends, but they did not act very interested. Will insisted that they come farther down the beach and have a look since it was impossible to view the lights when they were near the fire, because of the glare.

The lights were still hanging there when he pointed them out. One of his friends said, "Oh yeah, I see them." But the other thought it was probably a helicopter or maybe a tower with a light on top.

Will was used to studying things through his trained photographer's eye and this was interesting enough to photograph. He had a camera in his car, a Pentax loaded with slide film. He retrieved it and, walking some distance from the fire, looked for a rock to place it on, because he figured the lights would require a time exposure and he didn't have a tripod with him.

He had another problem as well. The Pentax had no built-in timer and was usually triggered by a squeeze bulb for time exposures. The only other alternative was to hold the shutter release button down with his finger, but this was risky because he planned to expose the image for a few minutes and would certainly shake the camera during such a lengthy period. He placed the Pentax on a rock, lined it up as best he could, placed a small rock on the shutter release, then left it there.

When Will returned about five minutes later and stopped the exposure, the lights were still hanging in the same position in the sky. He watched them for several more minutes, then went back to the fire. When he looked for them again a short time afterward, they had disappeared. The lights had remained in the same position in the sky for over fifteen minutes.

When the slides were developed, he examined the lights once more, pleased with the clarity of the pictures. He placed them in a safe place along with several thousand other slides he had taken, and forgot about them for many years.

11:00–11:30 P.M. ADT
Herring seiner fleet
Northwest of Brier Island and Digby Neck, Nova Scotia

Walter Titus was the master of the *Quadra Isle,* a herring seiner and part of a fleet of some fourteen or fifteen vessels plying the waters to the northeast of Digby Neck, Nova Scotia. The neck is a long thin finger of land, stretching out about thirty-four miles into the bay.

On the night in question the fishing fleet was strung out along the Digby Neck like a necklace of lights, chasing the herring up the Bay of Fundy. At night from the air or the water these fleets are a sight to behold. The very nature of their operation depends on

many high-wattage floodlights that light up the area
like a small city. It might have been this spectacle of
lights, shining so late into the night, that attracted the
attention of a visitor of a kind the one hundred and
fifty fishermen had never seen before.

Walter Titus first noticed something was wrong
when his crew stopped what they were doing and
started watching something in the sky to the southeast.
Before he could shout to them to get back to work, his
own attention was arrested by the antics of a brilliant
white light in the sky. He immediately discounted the
sight as anything of a normal nature. Like his crew he
was fascinated by what was happening in the sky.

The night sky was perfect for viewing, clear but
with no moon and festooned with stars. The light he
was watching was about the size of the moon and proceeded to give off three brilliant yellow lights that
formed a triangle around the larger light. The lights
slid effortlessly across the sky, then back again at great
speed.

Walter picked up the microphone on his marine
band radio to call one of the other vessels close by and
ask if they were seeing these things, then noticed that
the radio was alive with excited talk about the objects
in the sky. It became evident to the skipper that the
whole fleet had stopped to watch the sky, and little
work was being done. Down on the deck Walter's son
Bradford remembers seeing the objects far to the
south, diving toward the water.

Apparently some of those aboard the fishing boats
felt threatened by the lights. The Lent family, including

Mr. and Mrs. Lent and their three sons, ages thirteen to nineteen, were watching this show of UFOs from the shore with a few of their friends, using 7 x 50 binoculars, while listening in on the marine radio. Their next door neighbor, Albert Welch, was also watching with another pair of binoculars. One of the boats just off the coast of the island feared being crashed into, the Lents reported, and they overheard the fishermen say that the objects had four lights, each on some sort of an extension, each one flashing on and off.

Aboard the *Quadra Isle* the skipper heard an old friend of his, owner of his own boat, talking on the marine radio. The friend's name was Burton Small and his vessel was part of the armada of fishing vessels off Digby Neck that night. He was listening to his marine radio, and when he heard all the UFO talk, he went outside on the bridge of his boat with a pair of powerful binoculars that he had been issued during the Second World War. It was not long before he spotted the source of all of the excitement.

On the *Quadra Isle* Walter Titus heard Burton Small say to anyone listening on the marine band, "I wish you boys could see what I can make out with my binoculars." But he never explained exactly what it was that he saw. Walter always intended to ask Burton what he meant by that statement, but he never got around to it.

After observing the aerial display for about five or six minutes, Walter watched the lights depart to the southeast. To this day the memory of that brief encounter is still fresh in his mind.

As for his son, Bradford Titus, he would see another light later, during another encounter, blood red and streaking across the sky from horizon to horizon in seconds. Like his father he never forgot the sight.

11:00–11:10 P.M. ADT
Arthur Lake on Highway 304
Five miles southwest of Weymouth, Nova Scotia

Royal Mounted Police Constable Ian Andrew and game wardens Bert Green, Don Brown, and Sonny Wagner were on a stakeout in the forest near Arthur Lake. Earlier in the evening of October 4 they had driven into the woods to set up a watch station where they could observe trails and logging roads that were being used by deer poachers.

The night was cool and clear, and since these men were veterans at this job, they had come prepared with extra clothing, thermoses of hot coffee, and sandwiches. They staked out an area on the edge of a large clearing, overlooking some trails that they thought might produce results, and prepared for a long night. It was not uncommon for these stakeouts to go on into the early hours of the morning. They clipped off small branches that might obscure their view and used them as camouflage to conceal their location. They had their vehicle as a backup in case it got really cold.

The night was cool, about forty-five degrees Fahrenheit. There was no moon, but the sky was filled with stars that, although they were brilliant, offered little in the way of light. It was a good night for poaching. They waited, watched, but mostly they listened.

They listened for the sounds of distant vehicles that might be making their way up the old logging trails and roads to their position. They listened to the porcupines, skunks, and weasels rustling through the dead leaves that carpeted the floor of the forest at this time of year. These dead leaves acted as a natural warning system for deer, since they were difficult to walk through quietly. Hunters described it as like walking on Kellogg's Corn Flakes.

Contrary to popular belief the forest at night is anything but quiet. Even the deer themselves can be noisy as they travel through the woods. Owls and loons called out and the trees creaked with the slightest breeze.

The Mountie and the game wardens muttered to one another from time to time, swapping stories of earlier experiences, pausing occasionally to listen in the direction of some new sound. Constable Andrew used a small keychain flashlight and carefully checked the time. It was just before 11:00 P.M.

"Doesn't look like we're going to see much action tonight," he whispered.

"Could be the others will have more success farther south," Bert replied. As if to confirm that theory they heard a distant rifle report to the south.

They had gotten back in their vehicle by then. "Sounds as if you might be right," Don whispered back. "It's cooler than I thought it would be too."

The four of them whispered among themselves for a few moments. They had rolled down the windows so they would be able to hear better.

Gradually their conversation faded away, and they realized they were all staring at something through the windshield.

Finally Ian broke the silence. "What the hell is that?" he asked. He opened his door and got out of the car.

Bert, looking up as well, saw an orange-colored light, like a glowing ball of fire, south of them and just above the tree line, moving silently but slowly across the star field. There were spark-like objects emanating from it but there was no sound. "Do you see that?" he asked.

"Yeah, I see it. What is it? Is it an airliner?"

"If it is, it's the quietest airliner I've ever seen. Besides, it doesn't look right." Constable Andrew saw it as a candle-flame-shaped object but it looked as though the flame was upside down. He, too, remembers sparks and a corona around the object.

They watched for a few moments, standing on the logging road in silence as the object slid across the night sky, at what the Mountie guessed was a low altitude of about two to three hundred feet. Bert pulled a compass from his pocket and took a bearing. The object glittered and sparkled and then finally disappeared over the tree line at the south end of the clearing.

They spent the rest of their watch wondering what it could have been. When asked why he had not filed a report later, when he returned to his post, Ian Andrew explained that in those days, when they filed reports on UFOs, there was a tremendous amount of paperwork, with three different reports required by RCMP headquarters, the Air Desk in Ottawa, and the National Research Council. His caseload was heavy enough without adding to it. But the event impressed him enough that he could remember the details of it clearly twenty-nine years later.

C H A P T E R   T H R E E

THE SHAG HARBOR INCIDENT

Wednesday, October 4, 1967, approximately 11:20 P.M.
Atlantic daylight time
Highway 3, Lower Woods Harbor
Half mile west of Shag Harbor, Nova Scotia

The Dark Object now rocked gently on the salt waters of the southern passage. It drifted with the ebbtide, slowly, intelligently, toward the south end of the sound, and the deeper waters of the North Atlantic. It formed a dark silhouette in contrast to the warm glow of the lighted houses on shore.

Earlier it had caused concern for hundreds of people, including watchers in hidden places. Coded warnings about it had flashed through a secret network of warrens and enclaves to places where decisions could be made, decisions that would put in motion terrible engines of destruction. But it had done the one thing that it could have done to avert such a response: it had hovered, rather than attacked. Then it had seemed to disappear.

Now it drifted innocently on the water, not knowing the confusion it would create in the near future, the lives it would affect and change.

Only minutes before, Laurie Wickens, age eighteen, had been been driving his car along Highway 3, taking some of his friends home, laughing and carrying on with them.

It was getting late, nearly 11:25 in the evening. They entered the village of Shag Harbor from the east and Laurie slowed down, worried that a policeman might be parked off the edge of the road anywhere through the village. Almost immediately he saw an object flying low in the sky, flashing four lights, one after the other, in a straight line. He called it to his passengers' attention.

The girls looked up and had no trouble seeing it. It appeared to be descending at about a forty-five-degree angle, but not with any great speed. Sometimes it would appear to stop its descent and hover, then continue downward again. From his vantage point Laurie figured its angle and direction would take it into the harbor if it didn't pull up.

He did his best to keep it in sight, all the while taking care not to go off the narrow road winding through the sleeping fishing village. The waters of the harbor were a hundred feet away to his left. The wharf was surrounded with fishing boats, bobbing placidly at anchor. As he passed the fish plant on his left, Laurie was vaguely aware that he was now driving faster, but his goal was not to lose visual contact with the object, because he was sure beyond all doubt that this was an airplane that was going to hit the waters somewhere near the harbor. His passengers were excited now, sensing the seriousness of the situation, urging him to keep it in sight.

Laurie was doing his best, but the object wasn't making it easy for him. Neither was Shag Harbor, with all its little hills, winding roads, and houses and buildings clinging to the edge of the road. Now the object was so close that the slope of the car roof blocked their view. Laurie knew that if he could only keep it in sight, in a few moments the scenery would open up and the ocean would come right up to the edge of the road.

He steered around a turn, risking a glance upward as he saw it disappearing behind the tree line between the highway and the waters. He thought he heard a whistling noise, then a whoosh and a bang. One of the girls also heard a whooshing sound.

The car rounded a turn, climbing toward a small rise that brought them to a clear view of the waters of the sound. The object, whatever it was, had just impacted the surface of the ocean about two to three hundred yards offshore. For a few minutes the five of them watched the Dark Object drifting on the surface, showing a pale yellow light, bobbing about on the water. Then their curiosity and sense of wonder began to change to concern that the object might be a crashed airliner. They decided to report it to the police. Laurie got back in the car, pulled out onto the road, and sped down the road two miles to the village of Woods Harbor and the only pay phone in the area.

At the Royal Canadian Mounted Police Detachment in Barrington Passage, Corporal Victor Werbicki had been finishing up some reports and manning the phones. This late in the evening there was no one there to help him. As senior officer in the detachment, he worked long hours and it was expected he would take up the slack, since there was no dispatcher in the evenings.

The phone rang at about twenty-five minutes after eleven on that Wednesday night. Werbicki picked it up on the second ring. On the other end was an excited fisherman known to him as Laurie Wickens. Laurie advised the corporal that he was sure that an airplane of some size had crashed into the sound by Shag Harbor.

"Are you sure about this? You haven't been drinking, have you?"

Laurie denied this and said they had seen the lights descending toward the sound and that they had heard a whooshing noise and a bang. The corporal, still unconvinced, told Laurie to stay by the pay phone after getting the number.

Laurie was left standing in a pool of light cast by a solitary light bulb in the booth of the pay phone at a gas station nearby. He stared with anger at the telephone receiver in his hand, as a dial tone replaced the policeman's voice on the other end. He hung up the receiver. *What the hell's going on here?* he thought.

A female voice called from the car behind him. "What did they say, Laurie? Are they coming?"

Laurie walked back to the car parked on the edge of the pavement. "I don't know. They took the phone number but I don't think they believed me. They asked me if I'd been drinking."

"What!" The girl was surprised. "Did you tell them something crashed into the bay?"

"Yeah, I told them." Laurie was a little put out. But for the moment he hated to leave this roadside location.

Now that they'd hung up on him, he didn't know what he should do next.

Meanwhile another line was ringing on Werbicki's phone. He pushed a button to connect him and identified himself. This time a woman's voice came over the line, Mary Banks from Maggie Garrons Point, an area overlooking the sound and Shag Harbor. She informed Werbicki that she had heard a whistling noise and a bang and saw something out on the sound and thought an airplane might have crashed there. He thanked her and got her phone number and hung up.

Again the phone rang and this time another woman informed him that she and a friend were driving south near the shore on Cape Sable Island, which is about thirteen miles northeast of Shag Harbor. She told the corporal they had observed an object, flashing four lights in a sequence, descending into the Shag Harbor area and they thought it might have crashed there.

Another call came in from Bear Point just outside of Shag Harbor to the northeast, a man this time, mentioning a whistling noise, a flash of light, and a loud bang, indicating to him that something might have crashed somewhere near the harbor. Werbicki took down the details, thanked the man, and hung up.

Laurie considered calling the Royal Mounted Police detachment back again, but then the anger returned. Maybe he would go back to Shag Harbor and get one of the boats and go out himself and see if he could help. He opened the car door and slid into the driver's seat. As he reached for the ignition key, the pay phone rang. He glanced at his girlfriend, and the phone rang again.

Laurie got out of the car, catching it on the third ring. "Hello?"

"That you, Laurie?" a male voice asked.

"Yeah."

"Good, I'm glad you're still there. It's Corporal Werbicki at the Barrington Police Detachment. I'm sorry if I was a little skeptical earlier; however, we've since had several calls reporting lights or an airplane going down in Shag Harbor. I'm dispatching two constables to that location and I'm heading over there myself right now, so would you go back to the Moss plant and wait for me?"

Laurie agreed to do so. "There's four other people here with me. They'll be coming along."

"Sure, fine." The corporal paused for a moment, then asked him, "Do you have any idea what it is?"

Initially Laurie had been inclined to think it was the lights of an airliner, but now, when he recalled the image of the glow on the water, he was filled with doubt. "At first I thought I did, but now I'm not so sure."

He got back in his car and headed for Shag Harbor.

There was a lull for the moment, so Werbicki took the opportunity to make a radio call to the only two officers in the area still on duty—constables Ron O'Brien and Ron Pond. Pond and O'Brien had been working the area southwest of Shag Harbor because of illegal poaching activity that had been reported in that area. Now they were making their way back east, on Highway 3, going right through Shag Harbor at just about the time of the occurrence. They were about four miles east of the area when they received the radio call from Corporal Werbicki ordering them to return to the detachment ASAP. They were tired after a long day and an even longer evening, but nevertheless they proceeded to report in at the Barrington detachment. For them that night was going to get even longer.

When he had finished telling them to meet him and Wickens at the Irish Moss plant, Werbicki hung up and immediately made a call to the RCMP Subdivision in Halifax and reported his concerns to the duty NCO who picked up the phone there. Halifax in turn passed the information on to the national headquarters in Ottawa. From there a UFO report was made to the air force in the same city and it was that agency that called the Rescue Coordination Center, or RCC for short, in Halifax, Nova Scotia. These were the men who would attempt to rescue passengers from a downed plane, if there really was one.

When Werbicki raised the duty NCO at the Halifax Subdivision, he expressed concern about the possibility of an airplane crash in his vicinity and requested a check as to whether they had any missing aircraft— military, commercial, or private. The NCO replied that he would look into it. Werbicki said he was heading to the site and would call later to find out what they had discovered. The phones seemed to have settled down for the moment, so he locked the office and went out to his cruiser.

Constables Pond and O'Brien arrived at the station in time to find Corporal Werbicki in the parking lot standing next to a radio car, pulling on his jacket. He informed them with some degree of concern that there was the possibility that an airplane, possibly an airliner, had crashed into the sound adjacent to the southern approach into Shag Harbor. The two constables expressed surprise, as they had just passed through the area and noticed nothing unusual. They had probably passed Laurie Wickens while he was waiting at the phone booth for Werbicki to call him back.

At about the same time that Laurie Wickens spotted the UFO, Norm Smith and Dave Kendricks were returning home in Dave's 1962 Chevy from a date with their girlfriends on Cape Sable Island, fifteen miles northeast of Shag Harbor. At the time they were seventeen and eighteen years old, and since Wednesday was a school night, they had to make sure their girlfriends were home early. They were proceeding westward through the woods on Highway 3, about three miles east of Shag Harbor. They were talking about their girlfriends and about girls in general. In fact, some years later Dave would marry his girlfriend, but for now there was no talk of anything like that.

Highway 3 was, and still is, a relatively deserted stretch of narrow paved highway. It is bordered on both sides by stunted spruce trees and alders that are twisted by the relentless southwest winds that frequently blow onshore off the tempestuous North Atlantic. The road climbs and dips like a poorly designed roller coaster. The fog had rolled in, and the headlights glanced off vague, twisted shapes. In the early hours before dawn this road can be a very spooky place. When there is no moon, you can't see your hand in front of your face.

Dave Kendricks was steering the Chevy deftly over the small hills and around the curves, as he had many times before. His friend Norm Smith was on the opposite seat, next to the window. They were still talking about girls when Norm spotted the lights of an object over the trees toward Shag Harbor. What he assumed were lighted windows were pointed at a forty-five-degree angle to the ground. As many as five lights were glowing steadily, from a dull red to an orange color. What he found most strange was the fact that they did not appear to be moving. He brought them to Dave's attention, and Dave spotted them too.

Norm remarked that it looked like the four lights were heading downward into Shag Harbor, a couple of miles ahead of them. His first guess was that the lights were on an airliner but that it was awfully low. And it didn't look quite right. At this point he knew instinctively that there was something about this craft that wasn't normal. Norm was experiencing that special instinct reserved for those who have a UFO sighting. All of his past experiences with aerial phenomena flashed through his memory, as he attempted to find something that would explain what he was seeing. But nothing popped up.

Dave, Norm, and the residents of the area were not strangers to aircraft sightings. Since the village of Shag Harbor was only about twelve miles from the radar base at Baccaro, it was not unusual for air force pilots (Canadian or American) to try to penetrate the radar web and test its defenses using planes with supersonic capability. Most of these staged attacks would come from the sea, day or night, in an attempt to get under the radar. As Norm explained, sometimes you would be out on your boat pulling traps or running out trawl lines and one of these things would come out of nowhere and "scare the living shit out of you" when it went over only twenty feet above the boat. Once or twice they even came down the Clyde River at treetop level.

But he was sure this was something different. He encouraged Dave to keep it in sight. Then the lights dropped behind the trees, and it seemed they had finally lost sight of it. They speculated on what they had seen but could not figure out what it might have been.

They arrived in Lower Shag Harbor after driving for some minutes in silence. Dave dropped Norm off at his father's house and went straight home to bed. He mentioned the incident to his mother, then turned in.

He went to sleep immediately, not realizing what was about to unfold later that evening while he slept. But for Norm Smith this night was just beginning.

Norm had to get up early the next day. It was only Wednesday night and he would be going out on a fishing boat the next morning. After the Chevy pulled away, Norm began walking up the path to his father's house. He caught a glimpse of something out of the corner of his eye. Turning left, he looked upward to a line of trees west of the house and was astonished to see the four lights once more, only this time they seemed to be heading down into Shag Harbor.

Norm ran into the house and got his father, Wilfred, who was still up, brought him outside, and showed the object to him. His father was at a loss as to what it might be. Fearing that this might be an airliner crashing into the waters of the harbor, they decided to go down there.

Wilfred hurried back inside, and woke up his brother, Lawrence. Then Wilfred got dressed quickly, and met Norm by the truck. They were intending to take a run in to the harbor and see if the object had indeed crashed into the water. They backed down the driveway to the road but Wilfred had to apply his brakes when he saw the flashing red light of a police cruiser coming from the east. As Norm noted, it was going "like a bat out of hell," headed west toward Shag Harbor.

Wilfred wheeled out onto the road with the comment that "something must be happening there to get those boys all worked up." He followed the cruiser as best he could, making the trip to the Irish Moss plant in record time.

When they arrived they were surprised to discover that there were already at least a dozen people standing at the water's edge, staring out across the waters of the sound in the direction of Outer Island. Norm went over and stood beside Constable Ron O'Brien. Constable Ron Pond was standing beside him. He saw Laurie Wickens and some of his friends. Corporal Werbicki rolled up to the gravel lot of the Moss plant several minutes later.

Werbicki assessed the situation. The light was still bobbing around on the water and appeared to be part of a large object he guessed to be about sixty feet across. He asked some questions of the young people there, concerned that people might be dying out there on the sound. He put Constable O'Brien in charge of finding a boat and crew they could use to get out to the area.

The tide had crested at approximately 9:48 P.M., about one hour and forty minutes earlier, so the current flow was out to sea. Not long after that Constable Pond, who had been ordered to keep an eye on the light, watched it grow even fainter through his binoculars, then disappear from view, appearing to slip under the water.

There had been some discussion among the Mounties about the possibility of finding a dory and rowing out to the area in question, about one half mile distant. One of the young people on the dock mentioned that there were fishing boats tied up around Maggie Garrons

Point at the Government Wharf. Norm Smith had mentioned a local fisherman named Bradford Shand to Corporal Werbicki as a possible contact. With the situation now doubly urgent, Werbicki left the area to contact Shand.

Werbicki had visions of people drifting helplessly with the riptide that could run as fast as four to five knots in the sound, waters that were at the best of times bone-chillingly cold. Hypothermia was a deadly concern for those unfortunate enough to be immersed in the waters of the North Atlantic, particularly where there was a strong current running. They would soon be exhausted from attempts to fight the current, the cold sapping their energy and will to survive. This brings about a lethargic state and a sleepy feeling, with death occurring shortly thereafter.

The number of observers along the shore was slowly increasing as people driving by or living in the area came to the site, drawn there by the sight of the RCMP cruisers and the number of people standing around. Some of them assumed that there had been a marine mishap, a possibility that was a part of daily life in all fishing communities.

Before leaving, Corporal Werbicki told Constable O'Brien to call the RCC (Rescue Coordination Center) in Halifax and ask them to do an initial phone search in order to determine if any aircraft were missing or overdue, unaware that this was already being carried out in Ottawa. O'Brien had to go to a nearby house and use their phone. His phone call was recorded by the

RCC at 11:38 P.M. It would be early the next morning before the RCC put a call through to *Coast Guard Cutter 101,* tied up in Clark's Harbor about six miles away on Cape Sable Island, and sent it to the site. It would be nearly 12:30 P.M. the next day before the vessel appeared on the scene.

Bradford Shand was at home in bed when he received a call from Corporal Werbicki asking for his assistance to help locate possible survivors from a plane crash in the sound. He did not hesitate to offer himself and his boat. He also suggested that another boat might help and Werbicki agreed. When he hung up he called Lawrence Smith, Norm's uncle, at his home. They were friends and he knew he could rely on his judgment.

Both Shand and Smith had arrived back in port earlier that afternoon and were tied up alongside one another at what is called Government Wharf. Lawrence Smith's boat was tied up on the outside. It was common practice to do this and it was not unusual to see six or seven Cape Island fishing boats tied up side by side. They got maximum breakwater protection this way.

Lawrence, as Shand knew he would, said he would meet him at the boats as soon as he could get dressed and get down there. Lawrence met Shand in about ten minutes and by this time there were Mounties and volunteers waiting to assist them. They jumped in their boats and headed out to the channel and out into the sound.

On his boat the *Rhonda D,* Lawrence Smith figured that from the information given him as to the approximate position of the impact site, he should sail a course between his present position and the light buoy at the northern end of Outer Island. Allowing for the current flowing out through the channel, he should soon come upon the aircraft and any survivors. Not far behind the *Rhonda D,* Brad Shand's boat, *Joan Priscilla,* was making slightly to the south of him, so as to cover the area when they arrived at the site. They talked back and forth on their radios, their voices tense with expectation. They were, after all, expecting to find plane wreckage or, at worst, bodies from an airliner crash, drifting with the tide.

Lawrence's brother Wilfred had already switched on a powerful "seal beam" light and was shining it ahead of the *Rhonda D* on the smooth surface of the waters.

On the *Joan Priscilla* Norm Smith remembers the crew speaking in hushed tones, so as to maximize their chances of hearing cries for help from any survivors that might be out there. Constable O'Brien recalls looking to the stern of the boat and seeing a subdued young fisherman sitting on the transom, his arms folded, scanning first the port side, then to starboard.

Norm Smith was disturbed by the thought of what he might find over toward the island. He was not only concerned about the possibility of seeing human body parts floating in the water. There was something else bothering him too. He had been one of the witnesses who saw the plane descending toward the harbor, and he was not thoroughly convinced in his own mind that what he had seen was an airplane at all.

Meanwhile, Corporal Werbicki was braced against the wheelhouse, holding on to some rigging, scanning the waters ahead for any sign of survivors.

They were making about ten knots. The diesel muttered belowdeck, its muffler growling from above the deckhouse. The bow waves rolled back along the *Joan Priscilla*'s rounded hull, curling into little crests, hissing when they collapsed back onto the surface of the water.

O'Brien watched the *Rhonda D*'s light play across the water only one hundred feet north of them. Both vessels were nearly abreast when they encountered thick yellow foam near the impact site.

Lawrence Smith pulled the throttle back and slowed the *Rhonda D* to a crawl. *What the hell is this?* he wondered. Off to his left, Brad Shand had done likewise, his boat coasting into the foam. Lawrence called across the space between the two boats.

"What do you make of this, Brad?" In the cool, still night air, his voice carried easily, even over the noise of the boat engines.

Shand surveyed the waters around him. They were in a long, wide patch of glittering yellow foam, about three inches thick, that floated like shaving cream on the sound. The stuff was roiled by bubbles rising to the surface.

"Never seen anything like this before," he hollered back. "Can't say as I care for it."

Me neither, Lawrence thought. They were used to

sea foam, which would have been nothing new, but none of the fishermen equated this stuff with sea foam. In an interview nearly thirty years later Ron O'Brien, who was retired by that time, remembered the reluctance of the fishermen to sail through the foam.

Aboard Brad Shand's boat Constable O'Brien watched one of the younger men reach over the back of the boat and dip his hands into the foam. When he pulled them back in, his arms were covered by some type of oily substance. It didn't cause him any pain when he touched it. Everyone smelled it and felt its texture. It was neither hot nor cool. They all agreed it was not fuel oil or engine oil or anything they could identify. The Mountie was at a loss as to what it could be.

Afterward the fishermen would say that they didn't like having to sail through the stuff. Lawrence said, "If I was going out to fish, I would go out of my way to avoid it. But since we were looking for possible survivors we didn't have much choice."

Bradford Shand, a veteran fisherman and seaman, was interviewed shortly after the event by Michael James, a reporter for the tabloid newspaper the *National Enquirer*. He is quoted as saying, "I've passed that same stretch of water many, many times on my way to the fishing grounds, but I've never seen water like that. A patch of about eighty feet around was bubbling and covered in brackish foam. It was weird."

Norm Smith figured the slick to be about eighty feet wide and a half mile long, going by the length of Shand's boat. Lawrence Smith noted that the bubbles coming to the surface near the area made it seem as if something had sunk there.

He remembers a distinct smell of sulfur in the area near the rising bubbles and made a point of steering his boat clear of it. Since there was a chance some type of gas vapor was present, he was concerned about a possible fire hazard.

The fishermen and their crew of would-be rescuers began to crisscross the area in a random pattern in an attempt to locate any survivors. By this time four more fishing boats had come out to join in the search. They moved steadily toward the southern end of the sound to compensate for the tidal drift. They knew that if anything made it out into the riptide, there was little chance of finding it during the night. It would then be under the influence of the tremendous currents of the North Atlantic.

By 12:30 A.M. *Coast Guard Cutter 101* had arrived and joined the fishermen in their search for the crashed plane. By this time Lawrence Smith, having overheard some information from the Rescue Coordination Center in Halifax, wondered if they were really looking for a downed aircraft. More and more they were beginning to think of it in terms of a craft of unknown origins.

By 10:20 A.M. the RCC was referring to the object as a UFO, having eliminated the possibility that it was a crashed airplane. A message stamped PRIORITY from RCC Halifax to Canadian Forces Headquarters on October 5 outlined the report of the RCMP officials on scene. It describes the craft as a UFO and refers to it as the "DARK OBJECT," perhaps because its orange light was no longer visible beneath the water.

As curious as they were as to what had crashed into the sound, the men on the boats were getting tired. Some of them had been up continuously now for nearly twenty-four hours. At about 4:00 A.M., near daybreak on the fifth of October, Lawrence Smith and Bradford Shand turned back to the Government Wharf in the harbor while *Coast Guard Cutter 101* returned to its berth in Cape Sable Island.

They continued the search later the next morning, after getting some sleep and having a good breakfast, but their efforts turned up nothing. The following day, the sixth, the Royal Canadian Navy's Fleet Diving Unit arrived on the scene to begin preparations for a search that morning. It had taken a full day to get their equipment on-site from their base in Halifax, where it was stored on an aged destroyer permanently docked there.

Four divers began a search of the relatively shallow waters of the sound in an area suspected to be the impact point of the mysterious and elusive object of the evening of October 4. By this time the RCMP, RCC, and the Air Desk in Ottawa had, by the process of elimination, tagged it as a UFO.

The bottom of the sound was sandy and surprisingly uncluttered. The divers found that the visibility was about twenty feet, which isn't bad for diving in maritime waters. Unlike those down south, these waters are not populated by multicolored fish and eye-catching coral. Kelp beds and huge clumps of seaweed are common on the bottom.

The current in the sound was strong, often running at four or five knots, with the tide making it difficult for the divers to search properly. It was extremely tiring as well. The strongest of divers in light scuba gear can usually travel at the speed of only one knot, even while wearing flippers. Fighting four-knot currents must have been exhausting.

The divers began their initial dives from the deck of *Coast Guard Cutter 101,* but after some time switched to the more spacious rear deck of Bradford Shand's Cape Island boat, the *Joan Priscilla*. They searched for most of Friday the sixth with little success, at least none that was reported. Some of the local fishermen watched from their own vessels and at least one, Donald Nickerson, reported seeing the divers bringing up debris from the bottom. He describes it as being aluminum colored, but nothing officially indicates that this was the case. Mr. Nickerson said his impression was that the divers did not want them there and were reluctant to talk with them.

Milton Crowell, an engineer with Nova Scotia Light and Power, was on vacation with his family, driving around the shores of Nova Scotia after returning from a long trip to central Canada. While driving nearby, he had been getting news on his car radio about some mysterious object falling into Shag Harbor. He heard that the Coast Guard and the RCMP were searching the waters to determine what the object might have been. Since he was in the area, he made it a point to keep an eye open for what might turn out to be an interesting diversion.

Eventually the Crowells arrived in the area and found themselves on Highway 3 about midway between Woods Harbor and Shag Harbor, near the Irish Moss plant. Seeing several cars and a police cruiser parked there, Milton guessed that this was most likely the area in question. He pulled onto the gravel and parked. He and his son got out of the station wagon and walked over to the shore to watch a small duty boat as it patrolled the waters of the sound. They were both eating ice cream that Milton had bought a few minutes earlier.

With his son in tow Milton decided to go right down to the shoreline, so the two of them scrambled over large rocks to get to the water's edge. Once there, he decided to talk with some of the RCMP officers who were combing the shore for anything of interest. He approached them and asked what the divers and the boat hoped to find.

"Flying saucers," was their reply.

Milton was stunned by the candor of the officers' answer. He asked them if they were serious. Indeed they were, the Mounties informed him. Something unknown, classified as a UFO, had crashed into these waters and three of their fellow officers had seen it floating out on the sound.

Milton thanked them and left them to their search, while he continued up the shoreline. His son asked him if there really were men from space. Milton replied that if some strange little man did come ashore and ask him for his ice cream, he should give it to him.

Later Milton snapped a picture of his son standing

beside the family station wagon and of some other spectators who had come to watch the Mounties searching for pieces of a flying saucer.

The fact that the divers found garbage on the bottom of the sound, next to a busy fishing port, as Donald Nickerson has said, is not surprising. It would have been unusual if they had not. Still, it is strange that official reports did not state that the divers had brought some debris to the surface for examination, even if it was just sea trash. There are rumors that material of some type was recovered and shipped to Dartmouth, Nova Scotia, to the Naval Armament Depot there for processing. A confidential military source has confirmed this, which in turn is supported by the UFO organization APRO's preliminary investigation report which states, ". . . any recovered artifacts would be turned over to Mr. Maurice 'Mace' Coffey, the unit's scientific consultant. If anything of extreme interest was found, it would be turned over to the National Research Council."

The search continued until late on Sunday, the eighth. By Saturday the number of divers had increased to seven from the original four. On Monday they put their gear back into their truck and left the area.

On October 9 Canadian Forces Maritime Command officially called off the search for the UFO that either crashed or landed in the sound adjacent to Shag Harbor. Later evidence would suggest that the latter was the more likely the case. In any event, the official report would end with this statement: "Not a trace . . . not a clue . . . not a bit of anything."

There were no answers or attempts at explanations either. As puzzled as the Royal Canadian Navy and Air Force were, they apparently were not curious enough to seek solutions to an obviously solid UFO sighting—at least, that was the official line. Like the Roswell incident twenty years earlier, for all intents and purposes the Shag Harbor incident, like the Dark Object itself, had sunk into oblivion.

THE UFO CRASH THAT WASN'T

Early November 1967, 1:20 P.M. mountain standard time
The office of Dr. Norman E. Levine, University of Colorado

An unscheduled meeting took place between Dr. E. U. Condon, who was under government contract to study UFO phenomena, and Dr. Norman Levine, a scientist on Dr. Condon's committee.

Dr. Condon, noted physicist and close friend of Robert Oppenheimer, who had helped develop the atomic bomb, had made a special trip to Dr. Levine's office at the University of Colorado. He was concerned that he might have another unexplainable UFO case for his Project Bluebook.

He expressed his concerns about a UFO crash

reported in Nova Scotia. He was tired of getting calls about the case from the UFO organizations Aerial Phenomena Research Organization (APRO) and National Investigation Committee for Aerial Research (NICAP). The heads of those UFO organizations would give him no peace.

Levine informed Condon that he had spoken with someone at Maritime Command, the Canadian navy's eastern headquarters in Halifax. They felt that something had gone into the water but at that time had no idea what it was. A search had turned up nothing that was relevant. A call to RCMP headquarters in Ottawa indicated that they felt the search had been thorough and there would be no point to further investigation.

That pleased Dr. Condon, who figured that was enough to give him an excuse to close out this one. Dr. Levine informed Condon that Jim Lorenzen of APRO had called claiming they had received dozens of letters about the crash and that the Canadian navy was still searching out there. There were even newspaper stories. Some local residents said they had witnessed the recovery of debris. He thought another call to the Canadians might be in order.

Despite this Condon told him to class it as unexplained. Dr. Levine was upset by what seemed to be an arbitrary decision. There had been some disagreements on the committee over Dr. Condon's handling of the information on earlier cases. Accusations were being made that unscientific methods had been used to explain away what, in some cases, could not be explained. Some had gone so far as to resign from the committee, refusing to have their names connected to the report.

Condon, during a speech at the Corning Glass Works on January 25, 1967, stated that "it is my inclination right now to recommend that the government get out of this [UFO investigating] business. My attitude right now is that there is nothing to it. But I'm not supposed to reach a conclusion for another year." This did little to instill confidence in the UFO community that this would be a serious and unbiased study.

Dr. David R. Saunders, a highly placed and respected member of the committee, already believed the government was covering up the real evidence surrounding the UFO mystery. Condon's remarks at Corning only served to reinforce Saunders's suspicions that the study was being used as a way for the military to get out from under UFO investigations.

In this case, though, Condon's preordained conclusions would prevail for nearly thirty years.

There is no doubt that something extraordinary happened in Shag Harbor on that cool, clear evening in October of 1967. There was no doubt in the minds of those in the Canadian military, or of the residents and fishermen of that little village. No one in the American military doubted it either. Americans were keenly interested in UFOs spotted over Canada, because they might be missiles taking a stealthy path toward the U.S.

There was no doubt in the mind of Harold Shea, editor of one of the oldest and most respected newspapers in Canada, *The Halifax Chronicle Herald,* which ran the story on the front page with headlines two inches high—no small commitment for a newspaper as conservative as this one.

And there were the wire services that picked up the story and spread it around the world, reporting that a UFO had crashed and sunk in the narrow channel at the southern approach to Shag Harbor in Nova Scotia.

The Royal Canadian Navy sent their Fleet Diving Unit to the site in an attempt to find evidence of wreckage in the sound.

There was much speculation in the days and weeks that followed about what might have happened in Shag Harbor. But no answers were forthcoming—not from the government, the scientific community, or the military.

It's strange how this extraordinary incident slipped into obscurity after only a few weeks, and became ignored and forgotten. The details remained buried in musty archives and misty recollections, and would languish there for twenty-five years.

Twenty-five years later Chris Styles was watching the popular television program *Unsolved Mysteries*. "Perhaps we will never know what happened during those fateful days near Roswell, New Mexico, in July of 1947," Robert Stack told his audience of several million viewers, before his face dissolved to a commercial for a Toyota 4Runner.

Chris Styles pressed the off button on his remote control. He had been watching a rebroadcast of a segment of the now legendary UFO incident that took place in the New Mexico desert near Roswell Air

Force Base on July 4, 1947. The show featured many credible witnesses to this enduring mystery that had become a cornerstone of American ufology.

This segment of *Unsolved Mysteries* reminded Styles of a UFO incident that occurred when he was twelve, the crash of a UFO into the waters of Shag Harbor. That crash had personal significance because he remembered seeing it and trying to follow it on foot. It exploded into the press the next day.

He remembered one startling truth about this forgotten case. It had not been dismissed by the authorities. They had taken it just as seriously when the official search effort was finally ended as they had when it was first reported. No one ever disputed the fact that a sixty-foot craft slammed into the waters of Shag Harbor on the night of October 4, 1967.

So twenty-five years later, what had happened to this official mystery, the astonishing event that had once generated headlines and fed the press for weeks afterward? Where was its place in North American ufology? Why had it been ignored?

Chris decided to get involved. But where to begin? The answer was simple—go to someone who is an expert, someone who already knows how to investigate this sort of thing. Chris decided to call nuclear physicist Stanton Friedman, one of the world's most famous UFO researchers, who has spent thirty years researching the subject and has published many authoritative books in the field.

He remembered a local news story that had mentioned Friedman signing copies of his book *Crash at*

Corona at a bookstore in Halifax in 1992. The news item had mentioned that Friedman lived in Fredricton, New Brunswick—Nova Scotia's adjoining province.

Well, that should make it easy, Chris thought. A quick call to the long-distance operator and he had a phone number. He dialed it and got an answering service.

"If it's an important matter, sir, we have a number where he can be reached."

It was important to Chris. He got the number and dialed it without knowing its location. It was long distance, judging by the area code. The phone rang, as it turned out, in Friedman's motel room in Austin, Texas. At that moment he was preparing to leave for a lecture later in the evening.

"My name is Chris Styles. I'm calling from Halifax. I'm sorry to bother you there now, but I assumed that I was reaching you in Fredricton."

"No, that's okay. I have a few moments before I leave to give a lecture."

"Thank you. I'll get right to the point. I saw a rebroadcast of the Roswell segment on *Unsolved Mysteries,* and it jarred my memory of an incident that happened about twenty-five years ago, in 1967. Perhaps you might remember hearing about the crash of a UFO into the waters near Shag Harbor, Nova Scotia, around that time? It was quite a news item then. There was even a military search, as I recall."

"No, I can't say that I—oh, wait a minute. Is that near the Yarmouth end of Nova Scotia?"

"Yes. Shag Harbor is near the southwestern tip of the province."

Stanton paused for a moment, "You know, there have been times when people have approached me after one of my lectures in Nova Scotia and asked me if I knew of a UFO crash near there. Something about a UFO going into the water and turning it a strange color. I can't say I remember any more about it than that."

Chris was becoming excited. His memories were correct. "Yes, there was a great deal of yellow foam on the water's surface. I was an eyewitness, as a matter of fact. I'm not sure of the date. I think it was in the fall of sixty-seven."

"Chris, if you're going to pursue this case, I advise you to go to the newspaper archives in Halifax and find all the press clippings you can. This will give you a starting point. It might give you the names of some other witnesses as well, and if they're rural people, they might be easy to track down. They don't tend to move around a lot, especially fishermen. If you get any dates from the press clippings you might gain access to military files through the Access to Information Act."

"In Ottawa?"

"Yes—don't hold your breath, though. The government can be tightfisted with this type of information." Another pause. "Look, Chris, I really have to leave for that lecture but if you give me your address, I'll do some checking into it myself and let you know what I find."

Chris gave Friedman the address, thanked him,

and hung up. He was elated. Though he didn't know it then, he had just taken the first step on a long journey of discovery. He had begun the hunt for the truth about one of the most fascinating cases in ufology.

The chase had begun. The very next day Chris went to the main branch of the Halifax Regional Library, anxious to follow up on Stanton Friedman's advice. He needed a date for the crash, but rather than starting a long search through the newspaper files, he decided to follow a hunch. He went to the book section and found a copy of *Mysterious Canada,* a book about paranormal events, strange disappearances, sea monsters, and—UFO sightings. A quick check of the index and there it was: Shag Harbor. He flipped to the page and there was the date, October 4, 1967. But the information was sketchy, only one brief paragraph.

He went to the reference section and obtained the microfilm reels for *The Halifax Chronicle Herald,* a very conservative daily newspaper, with a long history stretching back over one hundred fifty years.

Chris was therefore shocked when the headlines jumped out at him from the front page on the viewing screen. Dated October 7, 1967, in letters two inches high, they read: COULD BE SOMETHING CONCRETE IN SHAG HARBOR UFO—RCAF.

A spokesman for a special and little known Royal Canadian Air Force department in Ottawa for the investigation of Unidentified Flying Objects said last night a series of bright lights glided into the ocean off Shag Harbor, Shelburne County . . .

Chris read the news report with a great deal of interest. Stanton Friedman had been right. There were references to many different people, chief among them Squadron Leader William Bain, the spokesman for the Air Desk in Ottawa. He also jotted down several of the witnesses' names: Corporal Werbicki and Constable Pond of the Royal Mounted Police, Lawrence Smith, and Laurie Wickens, names that would continue to pop up in police and military reports the more Chris dug into the facts. These were people we both would come to know personally over the years.

He now had names, but he was still hungry for more details. He read other references to the incident in the days following the crash in the *Chronicle Herald,* but could find nothing about it in the other local dailies or weeklies. The library staff suggested that he go to the Provincial Archives for coverage by other papers.

He followed this advice and presented himself at their offices the following day. Soon he was leafing through the weeklies, noting their spin on the story and looking for any detail that might have been missed by the *Herald.* Curiously enough, only one of the papers, the Shelburne *Coast Guard,* contained no information regarding the incident. Chris found this to be very strange, considering that the UFO had crashed practically in its own backyard. But as he would soon learn, in the UFO investigation business, whenever something didn't seem quite right, it was usually because someone linked to the investigation was covering things up.

Chris learned all he could from the newspaper

reports, photocopying as he went along. At the time he had no way of knowing that this trail would lead him all over Canada and the United States.

But on this day he went home happy, and even more curious and motivated than before. Now at least he had some names. He had the newspapers' version of the event and a rough idea about what had happened. But the story of the event was scattered among various periodicals, pieces were missing, tantalizing details and questions were left unresolved. One of these missing pieces was the possibility that the divers of the Canadian navy's Fleet Diving Unit might have recovered wreckage from the ocean floor.

He had an idea. He got hold of some old sounding charts for the area east of Outer Island, then bought copies of the 1992 current charts. His intention was to compare the charts of 1967 to present-day charts, in hopes of spotting some anomaly on the bottom that would show where the UFO debris might have been deposited. He was shocked to discover that this was going to be harder than he had imagined. Together with his friend Bob MacDonald he pored over hundreds of the older and newer charts of the area, comparing them.

Chris had had no idea of the number of soundings to be found in just one small area of one portion of the ocean floor. There were literally hundreds of them sprinkled over each chart. And then they discovered another problem. The earlier charts had been sounded by hand and each position had been plotted on the chart by the old loran navigation method. Its accuracy could be off by hundreds of feet on this scale. These measurements could also have been altered due to the fact that there was a highly secret submarine communications base at Government Point only thirty miles away.

The new charts, on the other hand, were extremely accurate. Because of this it was impossible to determine whether any of the older soundings on the bottom matched up with the newer plots. Chris and Bob spent hours trying to make sense out of the discrepancies between them. They were nearly cross-eyed when they finally gave up on the theory, but not on their desire to find out what might be on the bottom of that sound. It would be nearly three years before Chris would get a chance to look in earnest.

Chris was rapidly using up the small amount of information available in the libraries and archives in the Halifax area, but once more the Halifax Library Main Branch came to his aid. A source in the reference section suggested that he make an application to the National Archives in Ottawa for declassified UFO information under the Access to Information Act. He looked at a document explaining the act and found out that, after he made a formal application for such information, the National Archives had to respond within thirty days.

Chris had two paths open to him now. He could try to locate and set up interviews with some of the Shag Harbor witnesses, and he could request information from the government. The latter was relatively easy to accomplish, so he tackled that first. Besides, the information might develop into more leads.

He dashed off a letter on July 10, 1993, to the National Archives and was rewarded by a speedy reply in the mail on the twenty-eighth from Lana Merrifield, a research assistant there. She regretted that they had little on the Shag Harbor incident, but suggested that he might look for former military and police reports about it. She specified Record Group 77 at the National Archives. When Chris asked why she hadn't gone ahead and sent them to him, she informed him that this material was on microfilm and therefore was only available through the library loan system. They could be sent to an agency or library but not directly to an individual. He would have to request them from the library, and view them there. She also suggested that he contact the Department of National Defense.

Chris had wasted no time making a request to the Halifax City Regional Library for the RG-77 information, then called Stanton Friedman for some advice about asking the Department of Defense for any declassified documents they might have. Friedman is a mine of information when it comes to finding your way around the federal bureaucracies in both Canada and the United States. After some discussion Styles decided to call Defense Headquarters directly.

He received a telephone call back two days later from Lilliane Grantham, research assistant to the director of history. She was unfamiliar with the sightings at Shag Harbor. Chris gave her a brief history of the sighting, along with dates and locations, and she promised to get back to him with any information she could find, as soon as possible.

Like most of us Chris had little confidence in any government department getting back to him within a short amount of time, and therefore was pleasantly surprised when she called back two days later. Ms. Grantham had found some unclassified material which she informed him she was sending to him immediately.

By this time Chris had received a call from the Halifax City Library informing him that the RG-77 material had arrived from the National Archives. He was not sure what to expect when he presented himself to one of the reference librarians, gave his name, and waited while she looked through the boxes of requested items. Finally she came back to him and presented him with a small box.

"Five reels of microfilm." She pointed to some vacant viewers lined up along one of the walls. "You can use one of those to view them. If you want to make copies, you'll have to use one of the viewers with a copier attached."

Chris nodded. "Five reels of microfilm." He sat down at one of the machines and threaded the first reel of film. He glanced at a wall clock. It was 1:20 P.M. He would not leave until nine o'clock that evening, and even then he had just scratched the surface. There were literally thousands of UFO sightings listed on these films.

There were incidents from all over Canada, from 1947 to 1984. It should be mentioned here that while it might not appear that way, due to the higher profile that U.S. incidents receive in the press, the incidence of UFO sightings in Canada is slightly higher per capita than in the United States, and one third of those come from eastern Canada.

Once more he was joined by his friend Bob MacDonald, and the pair spent three days painstakingly going through the films, looking for something associated with Shag Harbor. Don Ledger examined these films as well, and attested to the fact that it wasn't easy going through all of them.

Chris and Bob were just deciding that perhaps the National Research Council had never had any files on Shag Harbor, or that what they did have was still classified, when they arrived at the microfilm T-1744. This one didn't look good either. Also, it was nearly time for lunch. Chris pushed the button and advanced to the next frame.

Bingo, there it was. A priority message to CAN-FORCEHED (Canadian Forces Headquarters) from RCC (Rescue Coordination Center) Halifax. The message advised that a UFO had impacted the waters of Shag Harbor. The report named RCMP Corporal Werecicky (it should have been spelled Werbicki) as one of several witnesses to the incident.

"Yes! Yes! Yes!" Chris raised his arms in the air. "Finally, something from the military!"

"I must admit I was beginning to wonder if they were even involved," Bob replied.

"Now we need the RCMP reports. Those guys were on site so they had to have made reports."

"We're getting closer, Bob." Chris finally felt encouraged.

They continued to sift through the information.
But although there were many documents alluding to
the incident, there were no reports by the RCMP officer in charge of the Barrington Passage detachment at
that time, Corporal Victor Werbicki.

Then they came upon a priority Telex from
CANMARCOM, an acronym for Canadian Maritime
Command, to CANCOMDIVELANT, another acronym, for the Fleet Diving Unit Atlantic. It gave instructions for the unit to task out of CFS Shelburne on
the HMCS *Granby,* proceed to Clark's Harbor, and
provide a diving officer and three divers to conduct a
search for an object reported by the RCMP in the
vicinity of latitude 43°30' north and longitude 63°45'
west. The object was reported as approximately three
hundred yards offshore.

The unit was to work in conjunction with an
RCMP officer and his detachment while being advised
by him as to the object's probable location. In the top
right-hand corner of the document was written "S/L
Bain"—meaning Squadron Leader Bain, head of the
Air Desk in Ottawa, which was then the clearinghouse
for UFO reports from civil and military agencies.
Printed in capital letters underlined three times was the
word *UFO*.

Slowly and painfully Chris was beginning to uncover what had happened on that cool, clear night in
October 1967.

FINDING THE WITNESSES

Chris wanted to talk to some of the divers involved in
the search of the bottom of the sound next to Shag
Harbor. He figured, why not go right to the people
who had firsthand knowledge about whether anything
had been recovered? This idea was a little naive, but in
those days he was full of hope and approached his job
with an open mind.

He enlisted the aid of an acquaintance of his father's, an instructor with the Fleet Diving Unit, Guy
Fenn. Fenn was able to give him the names of the
divers involved in the exercise and their addresses. Although some of these addresses were not up to date,
with a little detective work Chris managed to track
each of them down.

He made his initial contacts with the divers by phone. The first two were not interested in talking about their experiences, while another was in serious condition in the hospital with throat cancer and not expected to live. His fourth call was to "Harry," who was now retired and on pension from the navy, recently remarried, and working for one of the local institutions as a caretaker to supplement his income.

Chris spent some time telling Harry how he had located him, and mentioned a couple of people, including Guy Fenn, who might have been friends of his. He asked whether he could ask Harry some questions about Shag Harbor and the UFO incident.

At the time Harry had no problems taking to Chris about it, but he repeated the words, "Shag Harbor, Shag Harbor? No, I don't think I recall anything there, but I was on a couple of UFO operations with the navy."

Chris said, "Do you think you could talk about them?"

As interested as Chris was in hearing Harry's stories, he wanted to focus on Shag Harbor. But Harry insisted on referring to a second incident that had taken place just off of Shelburne Harbor. "You know about the one off Shelburne, don't you? You know there was no doubt about that one?"

Confused, Chris attempted to correct him. "You mean in Shag Harbor. That's where the recovery operation took place. And that is in Shelburne County."

"Well, yeah, maybe," Harry said reluctantly, "but I remember us being off Shelburne Harbor. We didn't go into the town or anything. I was there, and maybe it was Shag Harbor, but I don't think so. We were sending divers down in pairs and it was pretty hectic. There were several ships there for over a week." He mentioned bringing up debris and foam.

Chris was puzzled. Ships? As far as he knew, there was only a coast guard cutter, *CGC 101,* on the site. And the newspaper reports mentioned nothing about debris being brought up. The foam was another thing. The only foam that Chris was aware of was what the fishermen and the Mounties had referred to as yellow foam floating on the surface. Harry seemed to be talking about solid foam, or something of a structural nature, being brought up from the bottom.

Chris wanted to pursue this further, so he asked Harry if he would mind if he came over at a later date, to get further details and record the interview. Harry indicated that was fine with him. Eventually Chris set up an interview for April 9, 1993, Good Friday. Chris and Bob arrived at Harry's house with some recording equipment and were greeted by him at the door. Upon spotting the equipment he informed Chris that he had been talking with some of his former buddies and was now reluctant to have his story taped.

"I've always been a bit of a rebel," Harry said. "I haven't even told the family about this." Indeed there were quite a few of them there, waiting to hear him tell about the mission he was on twenty-five years ago. He even went so far as to suggest that maybe he should check with the navy to see if it was okay to talk about it. Chris quickly assured him that there was no need to tape the interview if he didn't want to. Harry showed them into the dining room. "Don't get me wrong. I said I'd give an interview and I will."

He invited Chris and Bob to sit down. He began to tell a little of his story, going over some of the details he had given Chris over the phone. Chris sensed that Harry wasn't as enthusiastic about parting with details as he had been over the phone, so he took out some press clippings he had brought along. This seemed to work. Harry chuckled over some of the details reported in the clippings, acting insulted at one point, where it mentioned that the divers were exhausted after searching all day. "Yeah, right, exhausted. That's a load of bull."

Harry mentioned that they had been bringing up a lot of foam. "Foam—you mean from under the water?" Chris asked. "The press mentioned a frothy foam on the water. The fishermen sailed through the stuff while they were looking for the object."

"No, we got big chunks of this stuff from the bottom." Harry spread his arms apart to indicate the size. "Maybe *foam* is the wrong word. Some of it was decomposing while we were bringing it up."

The ex-navy diver continued with more details, and again mentioned all the ships that were there. He also kept referring to the dive off Shelburne Harbor. None of this went along with the facts Chris had learned about the Shag Harbor UFO.

It was Bob who first ventured an opinion. "We're

not talking about the same thing here, are we? This is another incident. I mean, you're talking about several navy vessels sitting on top of this thing for a week."

At this point Harry realized that Chris and Bob had made the connection. Chris got the impression that Harry was divided—he wanted to reveal information that he knew should be kept secret. He now became reluctant to talk more about it.

Chris listened to Bob and Harry swap "war stories" for a while (Bob is retired from the Canadian army). One story Harry told was about the recovery of a body from an airplane, in three hundred feet of water off the coast of Maine. It must have reminded him of Shelburne, because he forgot himself for a moment and began describing some of the measurements they took with sonar and soundings to pinpoint the location of the object under the water.

Then he pulled up short. "I can't go any further with this," he said.

"Why not?" Chris asked.

"But we're talking about two different things here. You're the one who mentioned all the ships," Bob reminded him.

"Were the ships searching for an intact craft?" Chris was pushing him a bit now.

"Yeah, we knew there was something down there, and we knew it wasn't anything from here."

"What do you mean?"

Harry held up a hand. "I can't talk about this. I know you're not going to leave it alone, but I'm not going to say anything else. Look, we all saw it and there was no doubt that it wasn't anything from here. You know, I've dived on sunken Trackers and naval vessels. I've pulled bodies out of the ocean, and cut up material with torches underwater. This was the real thing."

After that he would say no more. Chris and Bob thanked him and said good-bye, then headed for Bob's car. Once inside, Bob turned to Chris. "You know, that's your real story. You should have pushed him harder."

"Maybe. But to be honest with you, I was a little confused. I'm not convinced there's anything to it." They argued about it on the way home.

Chris checked with some of his acquaintances, and even his father, who is retired from the navy, to see if anyone could shed some light on this latest information, with no success. He decided to try another call to Harry, on the off chance he could shake loose another piece of the puzzle. But it was obvious on the phone that he had lost him.

The ex-navy diver informed Chris that he didn't need any trouble. Chris tried several different angles, but they all just seemed to make Harry angry.

"You know you should drop it. You're never going to get the Shelburne story," he said.

At this time Chris had just received word that the Records Group 77 material had arrived at the library. "You know," he said, "I've got some records that have just come in from the federal government. Maybe I'll find something there."

"I don't care what you've got, Chris. You might

find out something about Shag Harbor, but you're never going to find anything in there on Shelburne."

"I've had people suggest to me that what you saw down there was wrongly identified as a UFO."

This made Harry furious. "I don't care what you think. I was down there! I don't know whether you're calling me a liar or an idiot, but you came to me, so I'm going to tell you something. I don't know what it was that was down there and I don't know where it came from. But it didn't come from this planet, I can tell you that! Now don't call back!" He slammed down the phone.

Several months later Chris was crossing the parking lot of the Sears department store in the West End Mall in Halifax when he bumped into Jim, an old friend of his from several years earlier. At one time they had worked together in the circulation department at one of the local newspapers. Jim had been retired from the Canadian air force for some years now. They exchanged greetings and Jim, who was accompanied by his young son, asked Chris what he was doing these days.

"You're not going to believe what I've been doing lately, Jim. Come on over to the car and I'll show you something." Chris had his backpack, which was filled with documents and newspaper clippings about Shag Harbor.

The minute Jim spotted the headline on the *Chronicle Herald* photocopy he said, "Oh, my God, how did you find out about that!?"

"Well, it wasn't that hard, Jim. It was in all the local papers at the time."

"Yes, I remember that now. You know about Shelburne, of course?"

Chris was stunned, but he managed to say, "Yes, of course."

"Yeah, they had all of those ships there. That was a hell of a mess," Jim remarked.

"How do you know about that, Jim?"

"I was there."

Chris admits he was more than a little surprised at this point. "Wait a minute now," he said, "you were in the air force. What were you doing on a naval vessel?"

"Good question. I wondered that myself at the time."

"What function did you serve? Why were you there?"

"I was there as part of an identification team. I'd been called in before to identify parts of aircraft, or the aircraft itself."

"What do you mean, aircraft parts?" Chris asked.

"Well, if for instance, a jet—say an F-104 Starfighter—went into the ocean or crashed in a forest up north and someone like a fisherman brought up some pieces of an aircraft in his net, or a trapper brought some pieces out of the woods years later, then they would bring me in to identify them.

"I even identified the nuts and bolts that went with certain aircraft. I had to identify a piece of wing off a Lockheed U-2 spy plane that the navy dredged up off

Newfoundland, back in the days when nobody had heard of them."

"Really?" Chris regarded him for a moment. "You used to mention a lot of things that happened to you in the air force when we worked together at the paper, but you never talked about this stuff."

Jim shrugged. "There's a lot I just didn't want to talk about, or couldn't."

"So what did they tell you they were looking for when you got to Shelburne?"

"I was told we were camped over a Russian sub that had made it in past the twelve-mile limit."

"Could it have been true? What made you think otherwise?"

"For one thing, my specialty is identifying aircraft. Why would they get me all the way out there to identify submarine parts? That made me suspicious. Also, we were kept belowdecks all of the time. And then there were the navy divers. They were constantly being reminded by the offices to keep their mouths shut about what they were doing below. They weren't even supposed to talk among themselves, but they did, especially at suppertime. The officers, some of whom were American, were really paranoid."

This was a new revelation for Chris. "American officers! What the hell were they doing there?"

"Don't forget, Chris, we were anchored right over the listening and transmission network for the biggest military submarine tracking system in the Atlantic Ocean. It was called MAD for 'magnetic anomaly detection,' which at that time was under American control. These devices, as well as God knows how many hydrophones, were planted all over the bottom of the ocean to detect Soviet subs. It was a highly sensitive and secret area. The Americans were bound to be there. They owned most of the stuff. They supplied Canada with high-tech equipment so they could find out if the Russians were heading their way via Canada."

"So what did these divers say to make you believe you were sitting over something other than a sub?"

"I guess the first indication was when I was listening to the divers talking over supper one night. And like I said, the brass was on top of these guys big time about talking shop, especially one diver." He mentioned the name of the diver we've called Harry. "The guy was a bit of a rebel. He'd had a few drinks and was saying stuff like 'Yes, sir, sorry, sir, we forgot about the sub, sir. I don't know about you guys but you know and I know that that ain't no sub. I don't know what it is or where it came from, but it ain't from Moscow!'

"So this officer tells him to settle down, but Harry's got plenty of booze in him now and he's argumentative. He's saying stuff like 'What are we, a bunch of kids? There's no reporters here. You know and I know that ain't no sub!' That's the kind of stuff I'd hear. They wouldn't talk directly to us about it but we'd overhear them talking about this thing down there. At first we thought they were full of shit, but when you saw the trouble they got into whenever they shot their mouths off about it, you had to wonder."

"And you didn't see anything yourself?"

Jim shook his head. "No. I don't know for sure what was down there. For a while I thought maybe they had lost an H-bomb out there somewhere, but that didn't seem to fit. Whatever it was, we were there for a week. Then one day we suddenly upped anchor and steamed out to sea at flank speed, and that was the last I heard of it."

Chris pressed Jim for more information, but he told Chris that was all he knew and seemed reluctant to go any further with it. He left, saying he had to get his son to hockey practice.

Now Chris found it even harder to ignore the story. Jim had raised even more questions that didn't fit in with the Shag Harbor information. And it didn't end there.

Chris had a chance meeting with a worker at a laundromat near his apartment. He was reading some press clippings while he waited for his laundry to dry, when a man who was working there noticed him, and said he remembered an incident that had occurred when he lived there as a boy.

"In Shag Harbor?" Chris asked.

"No, near Shelburne, around Government Point. That's where I lived."

He told Chris how the military had attempted to block the road so you couldn't get out to Government Point. Apparently they were letting the locals through, but no one else. This interested Chris, because from that vantage point, you could look over a two-mile stretch of water to where the flotilla of ships would have been anchored.

Chris told him he had some interest in an incident regarding the navy search in that area. "You might want to talk to the lighthouse keepers at Cape Roseway Lighthouse. It looks right out over that area."

Chris was inclined to agree with the idea, but wasn't optimistic about getting any cooperation from the Coast Guard.

"It doesn't matter about them," he replied. "I can tell you who the lighthouse keepers were. My family knew all of them. There was one man named Brenton Reynolds and another, Harry Van Buskirk. There was another guy there as well, I think he did maintenance or something, Peter Tapper."

Sure enough, upon checking the phone books of the area later, Chris found both of them. He found Tapper, too, who, as it turned out, was married to an old friend of his. Each one remembered the incident, but none of them were on duty at the time. They all remembered another light keeper they were sure was on duty that night, named Barry Crowell. Brenton Reynolds made an attempt to track down Crowell, locating several people with that last name in the area, but none of them turned out to be the man they were looking for. It was not until three years later that Don tracked down the real Mr. Crowell in Shelburne. His story added even more mystery to the already puzzling Shelburne story.

Late in the spring of 1994 Chris was visiting a friend who lived on the northern peninsula of Halifax. The man living next door walked across the yard and introduced himself. We'll call him "Earl," although that's not his real name. He said he knew a friend of Chris's. Earl worked for the city of Halifax now but had been a weapons tech in the military for many years. He was aware that Chris was "deep into the UFO thing," although he didn't know about his current investigation. Earl mentioned that he was interested in the phenomena due to some experiences he'd had in the military.

Chris's interest was aroused enough for him to tell Earl about his efforts to uncover the details about the crash. Earl said he was aware of the event, and had been posted at Shelburne afterward, and had been briefed about it. Chris asked him if he could talk about it, and Earl agreed.

"I could look at what you've got and tell you what I know, and maybe fill in some gaps for you."

Chris welcomed the opportunity. They agreed to get together a few days later. They met again early in the morning over coffee, before Earl left for work. Earl had been assigned to the Canadian Forces Station, Barrington, on Baccaro Point as a weapons tech with a top secret clearance and a specialty in missile recognition.

Earl had a UFO sighting in 1970, while on Cape Sable Island, about seven miles west of Baccaro Radar Station. He observed several orange balls of light that flew off the water at night, an occurrence not unheard of in that area. Because the objects he sighted were so close to Baccaro Radar, he broached the subject to his commander, Colonel Rushton, inquiring if this was some new weapon he hadn't yet been briefed on. While he was at it, he asked about the Shag Harbor incident, which he'd heard about three years earlier.

Chris was aware that what Earl had to tell him was secondhand. This was the information that was given to Earl during the briefing he received from Rushton: NORAD tracked the object when it entered the earth's atmosphere, after half an orbit over Siberia to the east coast of Canada and the Shag Harbor area. The military knew full well at the time that no airliner, space junk, or military hardware had impacted the waters of the sound. They also knew that the object did not stay put but submerged, proceeded out to sea, then headed northeast up the coast, around Cape Sable Island, and eventually came to rest off the mouth of Shelburne Harbor over a magnetic anomaly detection grid feed to the supersecret Submarine Detection Base at Shelburne.

Chris's antennae were quivering. There was that Shelburne connection again. Earl was telling the same story that Chris had been hearing over and over again.

Earl went on to say that a flotilla of six or seven naval ships—both Canadian and American—were anchored over the object and another that had joined it, apparently to help the first one with repairs. Divers were sent down to observe the two objects and photograph them both remotely and manually. Hydrophones and other equipment were lowered over the side to study them. The data they received from these studies was classified.

The flotilla remained over the same area for seven days, then was hastily dispatched to challenge a Russian sub that was threatening to penetrate the twelve-mile limit off Shelburne.

That was the extent of Earl's information regarding Shag Harbor and Shelburne Harbor. For the first time Chris had information that linked the Shag Harbor incident with the peculiar events off Shelburne.

This brought Chris to another individual who supplied a piece to the ever-growing puzzle of Shag Harbor. Chris was introduced to a retired air force officer by a friend who knew of his interest in UFO phenomena, and the Shag Harbor story in particular. This individual was an ELINT (electronic intelligence) officer attached to the 405 Squadron out of RCAF, Greenwood, in Nova Scotia. We'll call him Terry.

At the time of the incident he was taking special training in Port Hawksbury, Nova Scotia. This was interrupted when he was ordered to fly special missions from the American coast of Maine to Shelburne. The Canadian air force was interacting with the air force of the American navy, and regular conventions involving Canadian/American border incursions were being set aside, which he found unusual.

Planes from both air units flew patterns up and down the coast, dropping sonar buoys. Tension was high, entailing long, tiring flights during the mission. They flew the new Argus airplane, which could stay in the air for fourteen hours continuously. After seven days the mission was stepped down and nothing more was said about it.

He remembered one incident during debriefing sessions between flights, when the crews were joking with each other. They were taken aside, given a severe dressing down, and admonished not to discuss dropping the sonar equipment into the area with anyone outside, including their friends in the mess. They were to fly the missions, file their reports, and keep their mouths shut. Terry remarked that he had never in his career seen a mission handled in this fashion, even though intercepting Russian subs was part of their normal routine.

Another one of those curious coincidences occurred during a telephone discussion with Terry, while Chris was getting more details regarding the mission. He mentioned that his wife had lived in Shelburne in October 1967 (she was not married to Terry then) and said she remembered the Shag Harbor incident. Her father had some involvement with the armed forces at the time, and she mentioned that they had quarantined Shelburne Base and blocked the road to Government Point. The military was checking cars as they went through. That was three times now that Chris had heard the same story.

MORE WITNESSES

A t this point in time there is no reason to doubt that
something mysterious was going on off Government Point near the town of Shelburne at the same
time as the event in Shag Harbor. All the evidence suggests that the two incidents were linked. The Shelburne part of the incident was buried, its document
trail obliterated, and its witnesses effectively silenced.
Was this because of the UFO connection or because of
the base's military role?

Something happened under the water within rifle
shot of one of the most secret bases in the NATO defense community, a base so secret and so obscure that
for years, people living close to the base did not suspect that it was anything but what the Canadian and

American military complex said it was—an oceano-
graphic research station, set up to study the bottom of
the North Atlantic. What made it secret was its classified sonar system, set up, in cooperation with the U.S.,
to search for Russian submarines sneaking into United
States waters via Canada.

It took a sex scandal to bring its true purpose to
light. On March 1, 1985, the afternoon edition of the
Chronicle Herald broke a news story about a group of
lesbians in the Canadian military. This group of five
were being discharged because they could be black-
mailed by enemy agents, due to their sexual prefer-
ences. It was revealed that they were operators of
highly classified equipment for specialized top secret
operations concerning NATO defense strategy in the
North Atlantic. They were working at a high security
base identified as the Canadian Forces Station, Shelburne.

The newspaper reported then that CFS, Shelburne,
was one of sixty-four listening stations in a worldwide
underwater submarine detection system called SOSUS
for Sonar Surveillance System. The system was used
to track Soviet, American, and NATO submarines
cruising in the North Atlantic at a range of two to three
thousand miles.

The hull, prop, and engine signatures, produced by
ships and submarines as they moved through the water
(which is an excellent medium for sound), were
recorded as each vessel passed close to the hydrophones
and compared with sounds that had been previously
recorded, of both friendly and unfriendly vessels. These

sound signatures were on file in a computer in Nevada. If the ship was unrecognized and on the surface, a NATO aircraft could be dispatched to fly over it and photograph it.

It would be safe to assume that the North Atlantic was heavily salted with these hydrophones. An additional device was the magnetic anomaly detection, or MAD, grid. This grid was laid out in the shallow waters of the continental shelf and extended all the way from Greenland southward to the Florida keys. Anything with a magnetic field that passed over it, that is, any vessel not made of wood or fiberglass, could be tracked to within several hundred yards, perhaps even closer than that.

Any surface vessel can be tracked these days from space by satellites, but in 1967 NATO did not have eyes in the skies. They had their hydrophones and their MAD grid and intelligence networks to work with, and it was all very secret. Above top secret. So you can imagine what happened when a UFO came along and parked on top of that precious multibillion-dollar grid within two miles of a top secret base for seven days.

Chris now realized that while the government might be hiding UFO secrets with regard to Shelburne, it was definitely hiding military secrets about the port. This would make it even harder than usual to do research on the Dark Object.

He knew he would have little luck in getting information out of Shelburne because of its top secret nature. However, he thought he might have more luck pursuing his investigation at Baccaro Station, a more relaxed community nearby. Chris had the names of some sources to check out there, who were witnesses to the event at Shag Harbor.

The first one he contacted, on May 1, 1993, was Major Victor Eldridge, retired and now a county councillor in the town of Yarmouth, Nova Scotia. At the time of the incident in Shag Harbor, Major Eldridge was the base administration officer at Baccaro, a NORAD facility near Shelburne.

Chris satisfied himself that he was talking to the right person and then identified himself. He told Eldridge he wanted to ask him a few questions about the UFO crash at Shag Harbor.

Eldridge responded with "The what?"

"The Shag Harbor UFO crash."

"What do you mean? When was that?"

"Well, that's how it became known. It was 1967, October fourth. I'm sure you would remember it. Everybody else I've spoken to down your way has never forgotten this incident."

"No, I don't remember it. You've got to remember there were a lot of crazy things going on back then."

Chris said, "Well, considering your position at the Baccaro base, which was one of the staging areas for the search effort, you must have known quite a bit about this. Maybe this will jog your memory." He read him an appeal he had written, urging the public to report UFOs to the military or the RCMP.

Eldridge said, "Look, I'd like to help you, but I really can't." He continued with a question of his own: "Mr. Styles, what are you trying to do with this?"

"I don't know. I just find it very interesting. I've only been working on it for a short time, and the more I learn, the more there seems to be. It seems to be a very unusual case."

Eldridge offered some advice. "I think you will find that there is not enough to this to support any journalistic treatment."

Chris was a bit surprised by this remark. "That's a pretty strong opinion for someone who has no recollection of the incident." There was only silence on the phone. Finally Eldridge said he was sorry, but he couldn't help.

Chris first contacted another of the witnesses, Dave Kendricks, by telephone in the fall of 1993. Dave had witnessed the object while it was in the air and, later, on the waters of the inner harbor. At that time he was only a young man, eighteen years old. He had been returning from a date in Clark's Harbor, a small fishing community some ten miles to the north, on a large promontory of land that juts out into the Atlantic and is joined to the mainland by a causeway.

It wasn't until a few years later, while interviewing Dave again, that Chris and I found out that Norm Smith had not only been with Dave in the car and witnessed the object, but had also spent that night searching with the fishermen on the waters of the sound. As a matter of fact, since Norm was the passenger in the car, he had had a better view of the object than Dave. Dave went to the phone to call him, and as luck would have it, Norm was available and as anxious to meet as we were, so Dave invited him over.

While we waited for Norm to arrive, we sat around the kitchen table in Dave's house, which had been built on top of a hill on the eastern edge of Shag Harbor village. It had a view of the harbor and the Government Wharf. Even now the jetty was crowded with fishing boats tied up alongside one another, five and six in a row.

We both felt at home. Both of our families can be traced back to people who worked on the sea. These people have a strong sense of community and are self-reliant. When you drive through these fishing communities, people will look to see if they know you and give you a wave even if they don't. But these people only *seem* old fashioned; in reality they are shrewd and hardworking businessmen. When many of us are switching off the late show on television, they are rolling out of bed and going down to their boats at 3:30 or 4:00 A.M.

Many of these villages have become retirement areas, bedroom communities, and tourist attractions. You can always spot the home of a fisherman, though, because it's surrounded by lobster pots, fiberglass fish containers, ropes and chains, net floats, and stacks of lobster-pot buoys.

Dave Kendricks began telling his story. "I'd been on Cape Island to see my girlfriend. We dropped the girls off around ten-thirty or quarter of eleven and headed back to the harbor. It was a school night for them, so we had orders to have them home early. We were in my old Chevy, driving down through Baffling Woods."

"Norm Smith was with you at the time?" Chris asked.

"Yeah, we were heading west on Highway Three when Norm said, 'Dave, what's that up in the sky?' I asked him where it was, and he pointed up to the right of us, as we drove along. I looked up and you could see four lights, reddish orange in color. They weren't red and they weren't orange, they were sort of a combination, and they were at a forty-five-degree angle from the ground. You could tell they were moving. I'd keep glancing back and forth at them as I drove, and we suddenly came to a corner. As we went around the corner, this thing had gone down low enough so that it was behind the trees. I can't tell you how far away it was, but it appeared to be a mile or two from us. And it looked like something fairly big. What it was I can't tell you."

"Did you end up down by the Moss plant?" Chris asked.

"No. I let Norm out at his house and came home. I told my mother about it, then I went to bed."

"You were living here in Shag Harbor at the time?" I asked.

"Yeah, just down the foot of the hill," Dave explained. "I didn't know any more than that until the next day when I went to work."

"So you never called Barrington Passage Mounted Police?"

"No."

"I know Laurie Wickens was the first one to call, and they had several other calls. Did Norm Smith call in?" Chris asked.

"I don't know."

"I think he did, according to some of the press clippings," Chris said, "because he was the one who reported the whistling sound."

Dave agreed. "I had people from the *Chronicle Herald* and the *Enquirer* talk to me about it later," he said.

Chris went to our car to get a copy of the old *National Enquirer* report that Dave had never seen. In the meantime Dave explained that he'd thought the lights were some low-flying aircraft that was being tested nearby.

Chris returned with the *National Enquirer* article and showed it to him. "You didn't tell the reporters you saw a UFO?" Chris asked. "That's one of the things that makes this case so interesting. Most people call up and say they saw a UFO, but here, it was the RCMP and the air force that were saying that. Civilians, like yourself, thought they were looking at a crashed airplane."

"Even Laurie reported an airplane had crashed into the water, and a couple of others did as well," I said. "So they weren't even trying to report a UFO."

We spent a little more time getting the direction of the UFO's path straight in our minds, and had Dave draw us a picture of the line of lights he had seen and their apparent angle of descent. We wanted to be absolutely sure that this was the object that was seen diving toward the sound adjacent to Shag Harbor.

Norm arrived within ten minutes of being called. He lived just a short distance up the road on Highway

3. We brought him into the picture and told him about Dave's impression of the UFO's direction of movement and angle of descent. "We were just wondering if there was anything you might have seen that Dave didn't. He says he saw the object for about five or ten seconds. You were coming up through Bear Point, or past Bear Point, and Dave said it was off to the right, heading away from you."

"From what I can remember," Norm said, "the lights were more straight ahead of us when we came around that turn, and they were right over the top of the tree line. They were on an angle going down like this." Norm slanted his hand down at a forty-five-degree angle toward the table. "And they were going down to the left."

I pointed to Dave's map. "If you were coming in this direction here, the lights were going that way?"— indicating a southeastern direction. Dave agreed that was the case.

"Were they flashing in sequence at that time?" Chris asked.

"No," Norm said.

"It was just a steady row of lights," said Dave.

"That's what Laurie claimed he saw, and he wasn't the only one," Chris told them.

We had a few more questions for Norm. I asked him when he heard about all the excitement down in the sound adjacent to the harbor, and we were both surprised by his answer.

"I was out on a boat that night," said Norm.

"So you were out with Lawrence Smith?" Chris asked.

"No, I went out with Brad Shand. He was David's uncle."

"So you were out with Brad Shand and Lawrence Smith was out there too," I said. "Did you see any foam?"

"Oh, yeah, there was no problem seeing that," said Norm.

Chris followed up a point about the foam. "According to the newspaper the next day, when they talked to Brad Shand, he said the area of foam was about eighty feet wide and a half mile long."

"Well, Brad's boat would have been thirty-eight feet long, wouldn't it, Dave? When we sailed through it, it was about two boat lengths wide, so it would have been about eighty feet wide. And coming down through the sound, it would have been around a half mile long."

"What were your feelings when you were out there that night?" I asked.

"Scared, because I didn't know what the hell it was. We didn't have a clue what it might have been, although we all thought it might have been a big plane that had come down. We didn't know what we were gonna see when we were out there."

"Right, you thought maybe you were going to see bodies drifting around out there," I offered.

Norm nodded. "There's all kinds of things that go through your mind."

Chris found the press clipping he was looking for. He read the part of it reporting that Brad Shand, in his boat the *Joan Priscilla,* had steamed through the foam, called "the mysterious substance" by the paper, about twenty-five minutes after the light had disappeared.

"Did this foam look different or funny to you at all?" I asked.

"It was yellow! Yellow foam," Norm answered.

"Really yellow?" I pressed.

"It was almost like it was a glittery, shiny foam, like it was shining on top of the water," Norm explained.

"So you had no trouble telling the difference between that and regular sea foam?"

"Oh, Jesus, no," Norm stated emphatically.

We talked about generalities for some minutes, then Norm wanted to return to the moments just after they spotted the UFO.

"I'll just go back to when Dave and I separated. I got out of the car at Lawrence's house. Dad was living in one end of the house, the eastern end, and that's where I was staying at the time. I didn't realize until I got out of the car that the lights were still there."

This was new. We expressed surprise.

"I went into the house and got Dad to come outside. He had a look, then ran in to wake up Lawrence. The lights were still there when he came out. We jumped in his car and drove up the road. Before we reached the end of the driveway, the Mounties went by."

"Coming from Barrington Passage?" I wanted to establish the direction.

"Yeah, Barrington Passage. Their light was flashing," Norm agreed.

"That would have been O'Brien and Werbicki," Chris said.

Norm agreed. Chris said there was another Mountie who came in from the other direction. His name was Ron Pond.

Norm and Lawrence followed them down the road, unsure whether the Mounties were going to a car accident or if they were investigating the strange object in the sky.

This small mystery ended a few minutes later, when they rounded a turn just past Prospect Point and came upon the Irish Moss processing plant on their left. Cars were parked and people were gathered in a gravel parking lot that ended at the water's edge overlooking the sound and the approach to Shag Harbor.

They were right behind the RCMP vehicle when they pulled off the road and onto the gravel lot. "The Mounties got out of their car and we were right behind them. We went right to the edge of the bank. I stood right alongside Mountie O'Brien. He was looking off at the water."

Norm wanted us to be absolutely sure we knew the location of one particular bell buoy, which marks the outer limits of a sandbar that extends to Inner Island. Chris and I were both aware of it, and had been told that it had been lit up on the night of the UFO crash.

"When we were looking out there, the thing that we saw that night was close by, but it wasn't the bell buoy. I can guarantee you that. What we saw that night on the water, we didn't watch it for very long before it disappeared. But the bell buoy light was still there after it was gone."

"Could you see anything under the light, or just the light itself?" I asked.

"Just the light itself. There had to be something there, underneath the light, because the way the light was on the water, it was like it was floating on top of the water."

Norm Smith agreed to take us to the area where he and Dave Kendricks had their sighting the night of October 4, 1967. We all got in the car, and he drove around the area for some time, while we got our bearings and fixed the locations on our mental maps so we could come back and check it out more thoroughly later.

Later we said good-bye and drove off into the night. We arrived back in Halifax at about four in the morning after a long night's drive in foggy weather that could best be described as spooky.

GOVERNMENT DOCUMENTS

Chris decided it was time to check out some of the government documents he had been told about. On October 16, 1994, he arrived in Ottawa and checked into a hotel downtown. It was only one block from the headquarters of the National Archives and six blocks from the headquarters of the Department of National Defense.

Ottawa, Canada's capital, is located on the Ottawa River in Ontario. Canada's government institutions are all here, housed in austere and dignified stone fortresses. The Houses of Parliament stand alone in the center, surrounded by expansive lawns. In spring and summer these lawns are dotted with thousands of tulips, donated by the Dutch, in recognition of Canada's help in World War II.

Ottawa can seem like a miniature London, with Beefeater guards patrolling in front of government buildings and a canal that cuts through the center of the city.

The morning after he arrived, Chris went straight to the National Archives, filled out papers, and got a pass that accorded him access to areas holding pertinent military files that could be read but not copied. These were the reason for his journey.

With that out of the way Chris hurried to the third floor and filled out another form. This time it was a request form for a file designated as Target Detection Search, Flying Saucers, General, 1950–67. However, he was informed that he was not allowed access to that file group. Perplexed, he pulled out a letter that he had received from the National Archives, telling him that the file was available for research purposes at their headquarters. It suggested that new additions to Records Groups 18 and 77 might contain information relevant to Shag Harbor and Shelburne, so he was anxious to take a look at them.

The desk staff made a few phone calls, then informed him that they could fulfill his request after all, but it would take about twenty-four hours, because that file group was kept in another record depot, ninety miles away. Chris was beginning to experience the endless bureaucratic problems that go along with researching government documents.

The problems continued on the following day,

Tuesday the eighteenth. Chris discovered that the file he had requested had arrived, but had been transferred by mistake across the street to the Access to Information Review Center. It turned out there would be another twenty-four-hour delay.

The next morning Chris was again informed by an apologetic staff member that the RG-24 documents were still being held up. Chris decided to go across the street to see if he could hurry up the process in person.

At the Review Center he was referred to Bob MacIntosh, who was in charge of Access Reviews of Military Information (Canadian and American) and International Affairs. They talked for about an hour and he told Chris, "Don't worry. Potential embarrassment of the government or the military is not sufficient grounds for them to deny you access to the documents." Chris thought that was a fairly provocative statement.

Chris gave MacIntosh an overview of the Shag Harbor/Shelburne situation. MacIntosh in turn told Chris that such important documents would most likely never have been sent to the National Archives to be preserved. Instead, they would have been allocated to that level of secrecy alluded to as "above top secret," which means, for all intents and purposes, that the material does not officially exist. This allows the military to bury something so deep that even elected government officials are denied access to it. The many sections, subsections, and clauses of Canada's Access to Information Act give little direction on how to get around this evasive maneuver.

MacIntosh ended their meeting and Chris returned to the National Archives main building to find the file waiting for him. He was shocked at the small size of it. It contained only about two dozen documents, and none of them contained any information relating to UFOs. In the bottom file, labeled "temporary" on a yellow slip of paper, was a statement that this file group had had significant deletions in July 1994.

The last time someone had reviewed this file had been in 1984, and this had been done by one O. M. Solandt from the Defense Research Board of Sightings. It seemed to Chris as if the file had been sanitized. Even documents Chris had obtained earlier were missing.

Chris wondered why, after ten years, the documents had been thinned out. Perhaps it had to do with his guest appearance a month earlier on a popular Canadian morning TV talk show, where he had discussed the Shag Harbor events in detail.

Further attempts to get information at the National Archives, and later that day at Canadian Forces Headquarters, were useless. But later that evening he was more successful. He located the person whose name was affixed to several of the documents dated on or around October 4, 1967—Squadron Leader William Bain, chief of the Air Desk in Ottawa, at that time the clearinghouse for UFO reports for the Royal Canadian Air Force and the federal government.

The next day, Thursday, October 20, Chris met with Mr. Bain in the lobby of his hotel for several hours. Chris described Bain, who had been retired for several years, as a no-nonsense type, very direct. His gaze was steady and unflinching.

In the years 1963 to 1967 there were several officers who manned the Royal Canadian Air Force's air desk in Ottawa. Squadron Leader William Bain was one of the two majors, a colonel, and support staff that monitored, and filed reports on, all UFO activity in Canada. They also coordinated research efforts for physical evidence, with permission from higher authorities. Bain served at this post during the Shag Harbor incident in 1967, and for some months in 1968. Later this responsibility was taken over by the National Research Council.

Chris asked Bain if he was aware of a recovery attempt of a UFO off Canadian Forces Station, Shelburne, at Government Point during the Shag Harbor event. Bain claimed no memory of that event, although he was well aware of the recovery efforts at Shag Harbor. Nor did he recall any rumors about it, although he did express the opinion that something of that magnitude would surely have leaked out over the years.

When Chris showed him some of the evidence he had obtained, Bain admitted that it was enough to make someone justifiably suspicious. He went on to say that this could have been coordinated, and perhaps withheld, by the navy and NORAD.

On the twenty-first Chris again went to the National Archives to salvage what he could from their record files. He pulled RG-77 from the files and took a seat at the microfilm viewer. He was rewarded with a few more documents alluding to Shag Harbor that had not been present in the file group he had received months earlier in Halifax, including the description of young Darrell Dorey's sighting with his family.

At about 8:30 P.M. on October 4, 1967, the night of the Dark Object's crash, twelve-year-old Darrell Dorey, his older sister, Annette, and their mother stood beside their two-story frame house near Mahone Bay in Nova Scotia, southwest of Halifax. They were transfixed by the strange light show in the night sky. They, too, saw the orange ball of light, trailed by several smaller lights.

While they watched, the large object began to slowly merge with one of the smaller ones to become a single object. It wasn't clear to them if this was the result of one passing behind the other. In any event, it suddenly winked out, leaving nothing behind to show that it had ever existed.

Mrs. Dorey began to move the children back into the house, since it was a school night, when Darrell shouted, pointing up at the sky. As if it were giving an encore performance, the tiny starlike object blinked back on and began to dart around the heavens in rapid and impossible maneuvers. The three watched in silence until the finale, when it made a quick exit over the tree line, toward the ocean.

Its acceleration curve outdid anything young Darrell had seen at any air show. Another thing Darrell found very curious was the lack of sound, since the speed of the object should have caused a sonic boom.

Later, in the privacy of his own room, Darrell decided to write about it. It was too important to put off.

He began, "Dear Commander of C.F.B. Greenwood, I am writing about the UFO. . . ." He included drawings of the object in his letter.

At the end of the day Chris returned to the hotel, not much the wiser for a day's work and ready to head back home.

It's like pulling teeth to get information and documents out of the government, despite the fact that all of this information is gathered, investigated, complied, and filed at taxpayers' expense. But the difficulty factor seems to escalate sharply when it comes to UFO material.

Since searches of the ocean floor seemed to have come up empty, we were curious about whether the Dark Object was moving under its own power on the water while the witnesses watched that evening. There was some confusion on this point, even though everyone agreed it was moving out to sea. Chris at first had assumed that it was just drifting along with the ever-present current in that area. Two things that might affect its motion, besides the current and its own capabilities, were the tide and the weather.

We contacted the Federal Department of Fisheries and Oceans and asked if they had records of the tides back that far. "No problem," the technician told us, but it would cost to run a computer program to dig out that information if we wanted a hard copy. We told him that just knowing the answer would be enough for now, and within minutes we had an answer. The tides in that area had been ebbing for about forty-five minutes. An ebbing tide and the usual four-knot current outflow meant that the UFO must have been motoring along under some sort of propulsion because the current was too weak to propel it forward.

But according to the witnesses, although it seemed to be moving seaward, it was moving very slowly, not at the fairly good clip of six knots. So some other factor must have been affecting its speed. We wondered if there had been an onshore wind that might have been retarding the object's progress seaward.

So we called Environment Canada, the federal agency responsible for tracking and disseminating weather data, to ask about the detailed weather for that evening. We already knew from several sources, the RCMP documents and the newspaper stories, that it had been a cool, clear night with no moon, but we knew little else. The gentleman whom we contacted by telephone told us that there was no problem, they had weather information going back to the late 1700s. The data was kept in large journals for each month. He went looking for it and came back several minutes later and informed me in a very puzzled tone that the journal for the month of October of 1967 was missing!

"What do you mean, missing?" we asked.

"Well, this is strange. When these journals are removed from the stacks, they are signed out as to where they are going to, and for a date that far back, I can't imagine who would want it."

"How about somebody in your immediate office, perhaps they have it out for some reason?"

"I thought of that. There's only four of us here and

no one seems to know anything about it. I did find the daily log, which gives just the barest of information. Did you want that?"

"Sure," we answered. He informed us that the night over most of Nova Scotia for that date, the Shag Harbor area included, was cool and clear, with a temperature of forty-two degrees Fahrenheit, and low humidity.

"What about the winds?"

"That would be in the detailed journal."

"Has this ever happened to you before?"

"Well, the journals have been taken out of the office for one reason or the other, but we always knew where they were going and when they would be back. This is definitely unusual. Did you want to give me your phone number? I'll call you when I track them down."

We did exchange numbers and names, but we never heard from him again. We ended up getting the information from newspapers in the archives and have determined that there was little or no wind that evening in the Shag Harbor area. But it took a lot of work to determine that one little fact.

There's no proof that the weather journal was missing because of anything that happened on October 4, 1967, in Shag Harbor. But it did make it just a little bit more difficult to nail down that one piece of information, and many investigators would have quit right there.

The investigation into the Shag Harbor incident has revealed on more than one occasion that there might have been document tampering at higher levels.

Chris's experience of trying to get coast guard

cutters' logs is a case in point. He was anxious to read these reports, to see if any mention was made of the divers searching for debris in Shelburne. After many requests the log he finally received is so amateurish that it seems like a forgery. The times are wrong. They are recorded in GMT or Greenwich mean time, but it was daylight saving time on October 4, 1967. The Dark Object crash is noted as happening at 2330 GMT, which would have made it 7:30 P.M. local time rather than the actual time of the incident, about 11:30 P.M.

The entries in the log were supposedly penned over a forty-eight-hour period, yet they're all in the same handwriting. It's doubtful that the same person would have been on duty for two continuous days and nights, without a break. The log we received was in an ordinary spiral binder, but our investigation has revealed that logs are normally kept in bound volumes with numbered pages and printed labeling at the top. This is done to avoid tampering with the logs and changing timelines.

Lighthouse logs, and coast guard vessel logs for that period have turned out to be missing. The Stroker report, recording the daily reports and routine at CFS, Barrington, at Baccaro, disappeared, then showed up two months later. In the report there is no mention of an aircraft incident having occurred on October 4, 1967, only a few miles to the southwest, despite this being part of the NORAD eastern chain of radar defense for North America. This is just not believable, but it would stop all but the most motivated newspaper reporters from investigating further.

Frustrated with dealing with government documents, we decided to turn to the newspapers, hoping for some new leads we could follow up. The first paper to pick up on the Shag Harbor story, probably on a tip from someone local, was the Shelburne *Coast Guard,* a weekly that came out every Thursday in the Shelburne area. It dedicated three or four lines to the story of an airplane crash into the waters near Shag Harbor. There was no byline to the story, which probably was written at the last minute on Wednesday evening, October 4, as the paper was being put to bed.

Most of the weekly papers in that area are still published on Thursday for some reason, so none of them really had a chance to immediately cover a major, late-breaking story of this type in their area. But strangely, the largest daily in eastern Canada, *The Halifax Chronicle Herald,* did not pick up on the story until Friday the sixth of October. The *Herald* is one of the oldest and most respected dailies in Canada and in North America. In those days it had a provincial circulation in the hundreds of thousands, which wasn't easy in a province with a population of only eight hundred thousand people.

The normally superconservative *Chronicle Herald* shocked everyone with banner headlines two inches high announcing that there might be something concrete about the alleged UFO crash into Shag Harbor, according to the Royal Canadian Air Force. This was totally out of character for the *Herald,* both then and now.

Chris found this very interesting, and made some

inquiries of Ray MacLeod, the reporter who wrote the story, and who still resides in the area. Ray had told Chris to check with the managing editor for the *Herald* at the time, Harold Shea, and get the story directly from him. When Chris called him, he discovered that before Shea printed the Shag Harbor story, he had recently converted from being a full-scale skeptic to a believer in the existence of UFOs.

Mr. Shea and his wife were buzzed by an object on their way back to their home in Chester, a picturesque little fishing village, which has become an expensive yachting community, about thirty minutes from Halifax. The sighting occurred a few months before the event in Shag Harbor and was still fresh in Shea's mind when the story broke. He made a judgment call, realizing he had a chance to focus attention on a phenomenon to which, until a couple of months ago, he had given little notice.

Within a few days the debunking began. Ray MacLeod was pulled off the Shag Harbor story, apparently due to complaints from the general public that the articles were scaring them or their kids. The story was passed to another reporter, David Bentley.

At this point the events in Shag Harbor began to be portrayed differently in the *Chronicle Herald* and the *Mail Star* (the *Chronicle*'s afternoon edition). The press turned to an astronomer to give them guidance, credibility, and a way out of the situation. After all, the *Herald,* from the outset, had given the impression that the crash of the UFO in Shag Harbor had been a real event, something to be taken very seriously. The banner headline they gave the story was a good indication of this.

David Bentley consulted with the resident astronomer at St. Mary's University in Halifax, who happened to be a Jesuit priest, Father Michael Burke-Gaffney. He gave the kind of dismissive reply that one would expect. A priest—someone who, like many of the clergy, would be likely to have a bias against the idea of extraterrestrial life—seems an inappropriate choice for a newspaper to make if it is truly looking for an objective, scientific opinion.

Bentley's byline appeared under the heading "Shelburne UFOs Come Under Attack." This negative tone was a marked change from the earlier story titles that seemed to be upbeat and hopeful.

The next day another Bentley story appeared entitled "Shelburne's UFO: Secret War Machine from U.S. Scientists." In it theories are put forward by Canadian scientists as to the origin of UFO sightings.

Early in 1993 Chris attempted to interview Bentley, who was still a journalist but no longer with the *Herald*. It took several months for Chris to catch Bentley at his typewriter. He remembered nothing that could shed any light on how he found the scientists he quoted, but he promised to look into it for Chris. However, it was his colleague at the magazine, Lyndon Watkins, who called saying he had checked with some military buddies who informed him that an early prototype of the F-117A Stealth fighter was responsible for the crash in Shag Harbor.

In 1967? The earliest prototype of the Stealth flew

eleven years later and it crashed on the floor of the
Nevada desert. Without at least five computers aboard
the aircraft it is impossible to fly, and this technology
was definitely not available in 1967. This is reminis-
cent of the recent air force explanation that the alien
bodies that were reportedly discovered in a crashed
UFO in Roswell, New Mexico, in 1947 were actually
crash dummies dropped from airplanes in a test that
took place five years later.

The Shelburne *Coast Guard,* the weekly paper that
carried the story on October 5, mentioned not a word
about Shag Harbor the next time it was published, a
week later, despite the fact that Shag Harbor is right on
their back doorstep. The *Vanguard,* published in the
town of Yarmouth, only thirty miles away, carried the
story the following week, as did other weeklies up and
down the south coast. But not the *Coast Guard.*

Although the Shag Harbor incident died in the local papers relatively soon after the event, it did not immediately disappear from the international arena. This
was primarily due to the attention given it by several
publications, notably *Fate* magazine and the *National
Enquirer.* Their treatments of the event were precise
and surprisingly unembellished, especially considering
the tabloid nature of these publications. As a matter of
fact, they later became a very useful source for tracking down witnesses and for quotes from witnesses who
had passed away by the time we began investigating.

The highly controversial *Condon Report* cataloged
the event in its publication as Case No. 34, classifying
it as one of their few unsolved UFO sightings. The

Condon Committee did practically no research of their own. Acting on the basis of a report from James Lorenzen at APRO, the investigator, Dr. Levine, made several calls to Maritime Command and the RCMP, but they then dropped the case, because, according to Dr. Levine, "No further investigation by the project was considered justifiable, particularly in view of the immediate and thorough search that had been carried out by the RCMP and the Maritime Command."

With some of the heavy scientific hitters on the Condon team, they could no doubt have ferreted out some details that have since been lost to the passage of time. Here was an excellent chance to get in on the ground floor of an event as it was unfolding, but they dropped the ball.

The UFO organizations of that time didn't do much better. APRO was the only one that did a civilian investigation of Shag Harbor. They were also responsible for at least getting the event recorded in the Condon Report.

In September 1995, after our Shag Harbor research became a public news story, Don was driving to work one morning with the radio tuned to a Canadian Broadcasting Corporation morning news show. The subject being discussed was the event in Shag Harbor. One of the guests was Ray MacLeod, the reporter who broke the story in *The Halifax Chronicle Herald*.

He began by saying that every five years or so someone calls him up and asks about what happened there. He even mentioned that he had heard that there was some guy in Halifax who had been investigating the incident. Don Connolly, the show's host, asked him, "Then this was just another of those lights-in-the-sky things?"

Ray said, "Well, yes, that's true, but there was more to it than that." He cautioned Connolly that this had been no ordinary incident, that there were three Mounties on the scene who witnessed the object on the water and that the navy showed up two days later and did an extensive search. It was heartening to hear the man verify facts about an incident that was then twenty-seven years old. The mystery of it still grabbed him, but his journalistic UFO distancing instincts were still keeping him at arm's length.

Recently Stanton Friedman sent us an e-mail, telling us he was going to appear on a TV program called *Jane Hawten Live*. About two thirds of the way through the show she trotted out her astronomer, who was ill equipped to handle questions on UFOs. His main comment was that he believed that all UFO reports could be explained by natural phenomena. We turned off the TV feeling frustrated and disappointed.

Then we were again alerted by Stanton about an upcoming show called *Maritime Noon*. Stanton was to appear, along with Dr. David Lane, the resident astronomer at St. Mary's University in Halifax (at least they no longer had a priest in this role).

Somewhere during the program Shag Harbor came up. Stanton mentioned that we were involved in the investigation and that we lived in the area. Dr. Lane said that he had looked into the event and found no documentation to support it. I wasn't listening to the show, but eventually Chris got wind of it from a friend who was watching and he called in.

Chris asked Dr. Lane about his remarks concerning the lack of documents or details on Shag Harbor. Lane, who at least has an interest in UFO phenomena, repeated his difficulty in locating any documents except for press clippings in newspaper archives. Chris remarked that he had uncovered many documents during his investigation.

Lane pressed Chris for his source. "Well, for starters," Chris informed him, "if you go down into the basement of the university you are associated with, you should find some right there, in the files of one of your predecessors, Father Burke-Gaffney. You might remember him; he was the founder of your astronomy program at St. Mary's. Most of the other documents came from the National Archives and are microfilm copies of the original documents held by the National Research Council."

Admittedly Chris was a little upset by the astronomer's remarks regarding Shag Harbor. It does no one any good to have an expert in one area give his opinions about another area of which he has little or no knowledge. Astronomers are not automatically experts about UFOs. However, many people in the media don't seem to understand this.

CHAPTER EIGHT

THE MILITARY THREAT

On October 4, 1967, when the Dark Object bored its way into Earth's atmosphere above Siberia in the early morning, it was tracked by NORAD's Mid-Canada Radar Line and the Pine Tree Radar Line network. At NORAD several things must have happened when the object was first picked up on radar coming from the direction of Russia. The West was in the midst of the paranoia of the Cold War, so we can imagine the shock and alarm that must have swept through NORAD and the Pentagon when this thing first appeared. It would have been read by radar as a single incoming target, descending rapidly from a high altitude, with a high infrared heat signature just like that of an intercontinental ballistic missile (ICBM).

At first it would have looked like it was headed for the New England area, possibly even Washington, D.C. Pease Air Force Base in New Hampshire, the Strategic Air Command's B-52 Base located at Loring, Maine, and the U.S. naval base in Norfolk, Virginia, were only a few of its possible destinations.

When it showed up on radar, American and NATO bases would have been put on alert. Long-range interceptor aircraft all over the U.S., Canada, and Europe would have been scrambled. Missile silos throughout the United States might have been notched up to a defense condition (DefCon) somewhere between two and three, on a scale of one to five, with one being the most serious "launch missiles" condition.

And this almost happened. A knowledgeable air force source informed me that during the Shag Harbor time frame, North Bay scrambled fighters to intercept what was originally thought to be an incoming missile. He was advised that the object was traveling at 7,500 miles per hour, which is about Mach 10. In 1967 we had no aircraft that could move that fast, and we still don't today. Suddenly it stopped and hovered for some moments, then continued on its course at 4,400 miles per hour before slowing to a moderate speed and impacting into Shag Harbor.

All over the continent military commanders must have been calling base personnel back to their posts. Scrambled phone conversations would have been flying around the world in an effort to bring everyone up to readiness. The President would have been contacted and made aware of the situation and the military's defense condition.

There is a protocol in place for this sort of thing. It had begun fifteen years earlier, after the famous Washington, D.C., saucer invasion in July of 1952. The CIA got involved after receiving several reports, one of which was submitted by Edward Tauss, acting chief of the Weapons and Equipment Division. He recommended that the CIA continue to monitor UFOs, although no concern or interest should be shown by the agency at the public level, so as not to give legitimacy to the phenomenon.

Essentially, the military, ATIC (Air Technical Intelligence Center), and the CIA treated UFO phenomena as a possible smokescreen created by the USSR to interfere with North American radar and communications prior to an attack. At the very least they might use UFO flaps as a cover for low-level bomber attacks on the members of the NATO alliance. Smaller aircraft would be dispatched in advance to create confusion, well-lit planes that could be mistaken for UFOs.

The proposed scenario was that UFO reports would begin flooding in to the authorities and the air force would investigate and then dismiss them. After that the USSR could launch its real attack of military aircraft, so that the second wave would not be taken seriously, at least until it was too late, thereby giving the USSR the advantage. As silly as this idea seems now, in the early fifties it was a very real concern.

This was taken so seriously, in fact, that in December of 1953 the Joint Chiefs of Staff took measures to plug the UFO information flow to the public. They implemented what was known as Joint Army Navy Air Force Publication (JANAP) 146, with the subhead "Canadian–United States Communications Instructions for Reporting Vital Intelligence Sightings." It made releasing any information to the public about a UFO a crime under the Espionage Act (the National Securities Act in Canada), punishable by a one- to ten-year prison term and a ten-thousand-dollar fine. The law also applied to commercial airline pilots. It remained in effect until 1969.

Although government agencies in those days put on the appearance of caring little about the UFO phenomenon, in reality it scared the hell out of them, because they didn't have a clue about what the objects were. Whether or not that is still true today, it was the situation on that evening in October 1967.

Not long after the Dark Object appeared, we know that the duty officer at the local military base got a call from the local RCMP detachment asking if they had picked up anything on their radar, because there were reports coming in from all over about an airplane crashing in Shag Harbor.

There's a chance that the duty officer decided that he could not legally admit having received information about a UFO, because he wouldn't want to start a chain of events that could lead all the way up to the Pentagon and lead to his early retirement from the services.

The RCMP report says, "Baccaro Radar—

Negative." We've been asking ourselves the question "How could that be?" ever since we started this investigation. An object that was witnessed by at least twelve people to have slowly descended downward to the place where it impacted the waters off Shag Harbor and then was seen drifting on the water by even more people, three of them Royal Canadian Mounted Policemen—how could such an object have been missed by Baccaro Radar? The Dark Object was only thirteen-and-a-half miles from the radar station at its closest point, and this was one of the most sophisticated arrays of scanning and radar equipment that the military of that time could devise.

Baccaro had officially become a NORAD long-range radar (LRR) facility two years earlier, in September of 1965, meaning it had control over all aircraft entering its airspace. As such it would have had computers capable of handling multiple data input from radar spread through the NORAD system. This enabled air controllers and interpreters to construct an overall air picture without being overloaded.

Baccaro was plugged into the most sophisticated defensive grid in the world, with literally thousands of technicians all over North America watching radar screens, communicating, and interpreting information. Buried in secret rooms in secret places, controlled by the military and the intelligence agencies, were the most powerful computers of the day, capable of assessing the information pouring into them from radar installations all over North America, and making predictions about this information. Predictions were important in the case of a hostile, incoming missile, since it would be necessary to know its point of impact. It could then be intercepted by long-range fighters and shot down over relatively uninhabited areas.

What would the Soviets have been doing all this time? They have radar, too, good radar, and they must have been concerned. When this unknown object flew over them, heading for northern Canada and all of those NORAD radar bases, it would have looked for all intents and purposes like a missile. Since its track was describing a line in from Soviet Siberia, it could have been mistaken by the U.S. for an incoming ICBM. The Americans and NATO could have gone on full alert and maybe launched long-range bombers back at them.

However, it stopped and hovered, and ICBMs don't hover. If it hadn't done this, the Dark Object could have started World War III.

The best intelligence we have on the object's movements from this point on is the declaration of the retired weapons technician from the Canadian army called Earl.

Earl was transferred to Canadian Forces Station, Barrington, at Baccaro early in the 1970s. At his previous posting he had been told about the events that had taken place at Baccaro and Government Point on October 4, 1967. He was intrigued but did not pursue it any further. However, his job in weapons development moved him around quite often and he eventually found himself transferred to Baccaro.

He had not been there long, when one evening, while on Cape Sable Island, he observed several large glowing orange balls as they wafted ashore and flew away. He was stunned by the incident and decided to approach the base commander, Colonel Calvin Rushton, about his sighting and the incident at Shag Harbor in the late sixties. The colonel put him off at the time and told him he would brief him at a later date. Earl had another opportunity to talk to Rushton during a training session, when the colonel informed him that a UFO had entered the earth's atmosphere over Siberia, after a half orbit around the planet, made its way to the Shag Harbor area, and splashed down.

Rushton told Earl that the object had drifted with the tide for a while, then submerged and made its way out to sea, where it was picked up on the hydrophones at Shelburne. It then turned northwest and made its way farther up the coast, still submerged, and finally settled to the bottom near the MAD grid some two to three miles offshore, in sight of Canadian Forces Station, Shelburne, at Government Point, NATO's top-secret submarine tracking base.

Earl, now fascinated, was also told that a flotilla of seven navy vessels was tasked to the area and stationed over the object for the next seven days. The colonel apparently had no knowledge about what the divers discovered below, or chose not to reveal the details if he did. At any rate, at the end of seven days some of the ships were ordered to give chase to what was presumed to be a Russian sub testing the old twelve-mile limit. During this period the object presumably made its escape by heading south, still submerged, then finally becoming airborne again and heading out over the Gulf of Maine.

Why would Colonel Rushton tell Earl all this when, as an employee at a NORAD radar base, he would certainly have been sworn to secrecy? We can never know the answer to that. Perhaps it's just too hard to keep secrets all the time. As researchers we're thankful that vows of secrecy are sometimes broken.

The actions of local authorities would seem to support this story. According to one witness, the employee of a laundromat Chris met while waiting for his washing to be done, the military police had blocked the road to the base for some time during that period, and in particular that portion of the road overlooking the waters toward Cape Roseway on McNutt Island, where the flotilla of naval vessels was supposedly anchored over the alien spacecraft. He was not the only one to come forth with this story.

A Royal Canadian Air Force (RCAF) Argus ELINT technician was dating a young woman in the Shelburne area at that time. They are now husband and wife. She and her husband remember the roadblocks, as well as MPs checking cars as they went through the area, although they gave no explanations as to why they were doing so.

There is more evidence of the roadblock from other people living in the area at the time, but they are unsure of the dates. One couple had a close encounter near the blocked-off area on a rocky roadway leading down to the shores of Shelburne Harbor. They were so traumatized by this event that even thirty years later they will not discuss it. A friend of theirs disclosed that the couple came upon four to five entities near the shore. Something transpired there that the couple still will not disclose. Very curious happenings were definitely going on around that base.

We talked earlier in this book about the adventures of the navy diver Harry, one of the few who would talk to us, and Jim, the aircraft recognition specialist, who quite accidentally came out with the same story and remembered overhearing one of Harry's conversations.

The Shelburne UFO flotilla has been referred to by four of our witnesses: Jim, Harry, the ELINT crewman, and Earl the weapons specialist. All four were experts in their fields and all of them confirmed details of the other three without realizing it or being prompted.

No conspiracy could be successful without a cover story. An odd event occurred at this time which suggests it's part of such a cover story. A few years back, while Chris was looking for evidence of the naval vessels, he came across a story in the Shelburne weekly newspaper of October 12, 1967, the *Coast Guard*. The headline read, "U.S. Barge at Shelburne with Atomic Furnaces." The story went as follows:

> A large American steel scow being towed by a tugboat of the Moran Transportation Company of New York was brought to Shelburne for repairs last Friday night with a cargo of two huge atomic furnaces, en route from Philadelphia, Pa., to an atomic power plant at Sodus near Rochester, NY. The units, which were loaded on lowbed trailers, were too high to pass through overland routes, necessitating transportation by water up the Atlantic coast and through the St.

Lawrence Seaway. During the trip up the coast, the huge barge somehow partially filled with water so that there was only three feet of freeboard. Workmen at Shelburne Industries, Ltd., worked all Friday night to get the vessel pumped out while three divers from Atlantic Divers, a Liverpool, Nova Scotia, concern, made underwater inspections and repairs to the hull. The barge is 160 feet long and 46 feet wide and is owned by the S. C. Loveland Company, Inc., in Philadelphia. The towboat is the *Ann Moran* and it left Shelburne with her valuable cargo at 11:50 A.M. last Saturday.

Atomic furnaces? It would be a good cover story. Anything atomic keeps away the curious. If you are as suspicious as Chris and I, you start to wonder about every detail of everything that took place in and around October 4, particularly near the town of Shelburne.

We suspected the barge was actually used in an attempt to recover an alien craft. If this was the case, how would the crew get it to the surface about eighty to one hundred feet above and what would they do with it once they got it to the surface? At this point we have no idea what kinds of vessels were standing over the UFOs, but we assume that they were smaller ships such as destroyers or naval supply vessels.

We know that the Dark Object was about sixty feet long, and we estimated its weight at about sixty tons. We based our estimate on measurements that have been made of soil depressions attributed to landings of UFOs that were approximately thirty feet long. The depth of these depressions indicates that some of these objects had to have weighed at least thirty tons. Doubling the length would not necessarily double the weight, but it's the only way we could think of to guess the weight of the object.

Sixty tons isn't heavy for an aircraft, when you consider the weights of some of our own military aircraft, like the Grumman F-14 Tomcat, featured in the movie *Top Gun*. It has a length of sixty-three feet and weighs seventy-four thousand pounds, or thirty-seven tons. Remake it as a round object and fill in all the gaps and it could easily run up to seventy or eighty tons. A B-52G Stratofortress, for instance, has a maximum gross takeoff weight of 244 tons.

The problem would be that a destroyer could not lift an aircraft that heavy and would not have the deck space to carry an object of the sort they were searching for in Shag Harbor, which would have been about sixty feet long.

To get one of these UFOs aboard a ship, if it weighed sixty or seventy tons, would require a boat with great lifting capability and stability and a deck capable of supporting such a heavy deadweight. If an aircraft carrier had been brought in, this would have been a big news story. Canada's one carrier, the *Bonaventure,* was just coming down the St. Lawrence from a refit at that time. At any rate, the mouth of Shelburne Harbor did not have the clearance required for one of these.

But a barge would fit the bill very nicely. The very nature of its construction makes it an ideal conveyor of awkward and heavy materials. Barges are normally low in the water, and they frequently carry large cranes and derricks for lifting.

Still, there's nothing suspicious about a harbor with a barge in it. Chris was looking at all the floating stock in the area at the time, searching for anything that might have supported the flotilla theory and the possibility of a UFO recovery, or at least an attempt at a recovery. A barge would definitely have been an asset at the time. What made Chris suspicious was the mention of atomic furnaces, which sound like something out of science fiction. And if you read the newspaper article again you notice something else—the speed at which the barge was repaired.

The paper states that the barge was only three feet above water when it was pulled into the harbor, the result of an obvious leak. Yet the next day the barge is en route once more and supposedly headed for the St. Lawrence River. Once it entered the harbor, divers were hired from Liverpool, equipment was brought to the scene, the leak source was determined, the barge was pumped out overnight, and the divers were in the water making repairs, presumably at night, so that this barge could go on its way by noon the following day.

Chris, who had been around ships all his life, thought the speed at which the repair was done a bit unbelievable. Living, as we do, in Halifax, an area that has been a home to shipbuilding for over two hundred years, one tends to learn a great deal about ship repair over the years. Diving is a tough business, particularly at night. The water would be cold and any welding or patching job would be hazardous and labor intensive. From its arrival Friday night after dark (it would have been dark by 6:30 P.M.) until its departure from Shelburne Harbor at 11:50 the next morning is about seventeen hours, and a lot of that time would have been taken up with docking and departure. There wouldn't have been much time left over to accomplish the repair.

We got another clue several months later, when I was doing a phone interview with our ELINT crewman about his role in the aerial search for the UFO. I asked him if, when they were flying over the area near Government Point, he noticed any ships grouped near the point. "Oh, yeah," he said, "they had quite a few ships anchored over this sunken object. They even had a barge towed up from down in the States, from around Norfolk, Virginia, in a big hurry, to put the thing on when they got it up." That statement surprised us both, coming as it did right out of the blue. It dropped nicely into the puzzle. Could we confirm that there had been a barge in the area and find out where it had come from? Yes, we could. Was the barge capable of doing the job required by the cover story in the newspaper? Yes, it was. Did the reason for the barge being in the area, as stated in the newspaper story, make sense? For the leak, yes, for the cargo, no. Had the "repair job" carried out on the barge been handled in a normal fashion? No, it had been done too quickly to be believable.

So there it is, another small piece to add to the puzzle of this intriguing story, the story that, for Chris and me, won't go away.

THE SEARCH

By the summer of 1995 one question troubled us more than any other about the Shag Harbor incident. Could there be debris, artifacts, or even evidence of seabed trauma waiting to be found beneath the waters that lie between the Dark Object's initial impact site and its last known surface position? Could it be that simple to find proof?

It is a well-recorded fact that the original military search claimed "nil results." But one should bear in mind that the search carried out by the Royal Canadian Navy's mobile Fleet Diving Unit, stationed on HMCS *Granby,* was decidedly a low-tech, low-priority effort. It consisted of pairs of divers, working through the daylight hours with handheld, underwater lights from October 6 until sundown on October 8.

Undoubtedly they could not justify the cost of doing more. Former RCAF Squadron Leader William Bain, an ex-staffer on the Air Desk, told us that money was always a concern. In an ideal world he would have preferred to have at least three Sikorsky "Sea King" helicopters at his disposal. However, UFOs were not a matter of high priority with the military brass or the bureaucrats in Ottawa. In retrospect it is more likely that these types of military assets were rerouted to the waters off Government Point.

It seemed to us that there were three possible scenarios that could have taken place that October evening in 1967. Strong tidal currents could have carried the UFO farther out to sea, where debris was likely never to be found. Or the object could have come to rest on the seabed outside of the divers' search area. A third possibility is that, after submerging, the UFO abandoned the area under its own power, moving into the Shelburne area.

No matter which of these theories is correct, a sixty-foot unidentified flying object may still be resting on the seabed off Nova Scotia's southwestern shore.

By the summer of 1995 there were two things to consider. First was the state of underwater technology. Side-scan sonar, magnetometers, and subbottom profilers are powerful new tools of detection. Navigating by GPS (global positioning satellites) guarantees that what you discover on the sea bottom can be found again, with precision, on another day. This is why so much long-buried treasure is now coming to light. We would have advantages not available to the navy divers twenty-eight years earlier, but only if we could assemble the modern technology.

On the cold winter's day of January 12, 1995, we were at the Bedford Institute of Oceanography at the northern end of Halifax Harbor, for a meeting with Gordon Fader, scientist and oceanographer.

On that day Gordon's colleague Bob Miller sat in on the meeting. He seemed to be fascinated by the events in Shag Harbor, and appeared to find it refreshing that there was a UFO case that had significant corroboration and document support. As Gordon put it, "It seems to resist evaporating as one gets closer to it—just the opposite of most of these cases."

Noon approached, so we moved our meeting to BIO's cafeteria. As we passed documents around the table, Bob's forehead furrowed in thought. He informed us that he might have sailed over or at least near the Shag Harbor impact site on a geoscience survey cruise in the summer of 1988, six years earlier. After that, lunch became a hurried affair and soon we were on our way to Bob's office in another wing of the institute.

Moments after we entered BIO's Atlantic Geoscience Center, Bob had retrieved the cruise report of a survey conducted in 1988, between June 5 and 17, on the Department of Fisheries research vessel *Navicula*. The report's soft blue cover had a pocket containing two charts that were the result of the *Navicula*'s efforts.

Bob removed and unfolded them while Gordon

Fader unrolled the actual side-scan sonar data that had generated the *Navicula*'s charts. The two scientists were calling out coordinates to one another while we read a statement detailing the purpose of the *Navicula*'s cruise.

Cruise Report 88–018 (B) Phase II: "The purpose of this cruise was to collect shallow seismic reflection, side-scan, and magnetic data in support of a program to map the superficial sediments and shallow bedrock geology of the nearshore area south of Pubnico Harbor between Cape Sable Island and from Cape St. Mary's to Yarmouth. These data and their interpretations will assist in the processes that have and are affecting the sediments."

Specific objectives of the cruise were to assess the mineral sediments in the area and get magnetic data, which would help to identify them, and also to collect samples. The scientists also aimed to make a map of the surface and subsurface features of the area.

When Gordon located the side-scan sonar rolls that matched the Shag Harbor area, we pointed out on the *Navicula*'s cruise chart the approximate site of the UFO's last known surface position. It appeared that the *Navicula*'s closest course approach took it to within a mile of where the Dark Object had disappeared.

Then some interesting data caught our attention. On the *Navicula* cruise chart at the point nearest the UFO's last known position was a feature on the seabed marked by sonar interpreters "May not be boulders."

The sonar rolls were examined and the exact location of these acoustic reflections was determined. The sonar techs had written various comments in the margins of the paper rolls such as "What the hell are these things?" but all that Gordon and Bob could discern that day was what the mysterious acoustic reflections were *not*. They were not the remains of a shipwreck, depressions from drill rigs, shell beds, or dredge spoils. Oh, yes—they were not boulders.

A photograph of these strange sonar returns shows four circular depressions approximately sixteen feet in diameter, arranged in a diamondlike pattern in an area on the bottom of the ocean where one would not expect to see such features. It served to fuel our hopes that physical evidence might yet be found.

Of course the big question was, where were we going to find the money to mount an expedition to reexamine the areas in question? Our first and most logical resource seemed to be the Fund for UFO Research (FUFOR). We had gotten some support once before in the early days of our research on Shag Harbor. They had helped underwrite the costs of running newspaper appeals and a trip to Ottawa to search for files that could not be sent to us on microfilm.

We decided to call Don Berliner, who was projects director there, and discuss the situation informally with him before making any formal application. He seemed to like the idea and thought the reasons for a second search of the area were sound. He advised us to put together a proposal and include the reasons for searching the seabed for physical evidence, as well as a budget that would detail all possible expenses. He also mentioned that the project would probably be handled by Bob Swiatek, who had experience with diving and survey work.

During the spring of 1995 we spent a great deal of time learning about the nuances of conducting underwater surveys. We made countless calls to Glenn Gilbert of Canadian Seabed Research, Terry Dyer of Deep Star Resources, and, of course, Gordon Fader at BIO. All of these men gave freely of their valuable time and advice.

On expert advice from Gordon Fader at BIO, Chris Styles proposed a four-day survey, composed of divers, underwater video, 500KHz side-scan sonar, a magnetometer, and a subbottom profiler, with a budget of about eleven thousand dollars. After we'd submitted the written proposal to the Fund for UFO Research, informal discussions with Don Berliner and Bob Swiatek led us to believe that things were moving along nicely. A letter dated May 16, 1995, from Bob Swiatek informed us that the Shag Harbor proposal was creating quite a bit of interest. It was further encouraging that the Fund for UFO Research was now working closely with the UFO organizations MUFON and CUFOS in a coalition. Now they could merge their financial resources and take on larger ventures.

As the spring moved into summer, it seemed as if a kind of momentum had taken over the Shag Harbor investigation. Michael Strainic, Canadian national director for MUFON, was to speak at the International Symposium in Seattle that July. He planned to use a good part of his lecture time to talk about Shag Harbor. This would be the first time that MUFON's American membership would be exposed to this information. He felt it was among the strongest material he would be presenting during his history of Canadian ufology.

Paramount Television's syndicated program *Sightings* expressed interest in doing a segment on Shag Harbor for its fall 1995 season. It would be scheduled for a summer shoot during the very week that our funding proposal was to go before the new UFO coalition.

Far away from Nova Scotia, though, events were conspiring against us. The coalition was to meet in Las Vegas after the MUFON Symposium in Seattle. Two members each from FUFOR, MUFON, and CUFOS would meet with philanthropist Robert Bigelow to try to get funding for their projects. FUFOR sent Bob Swiatek and Richard Hall, who were going to present the Shag Harbor proposal. But something went wrong in Las Vegas that day.

Shortly after the meeting began, a major disagreement surfaced between Bigelow and the other participants. Mr. Bigelow wanted veto power over the coalition's decisions in return for his continued financial support. No one was willing to agree to this arrangement, so the Bigelow Foundation withdrew its support. As a result our proposal was never put forward. The August 1995 edition of the MUFON *Journal* did a short write-up about the meeting, under the headline "Coalition Reorganized for Second Year."

We were very disappointed at this turn of events, although both Bob Swiatek and Don Berliner assured us that an alternative solution would be found. But this would take time, and winter was coming, when such a survey would be harder, if not impossible, and certainly much more expensive.

While things were grinding to a halt in Las Vegas, Hollywood seemed to be getting it together. Paramount Television's *Sightings* show decided to go ahead with a Shag Harbor segment for the fall season. They planned to do a two-day site shoot that would be a brief overview of this relatively unknown UFO crash.

The *Sightings* episode was filmed in July. Producer Phil Davis called to say that they were happy with the footage they had shot and that the editing was going well. He also asked if we had received any funding yet, and learned about the disappointing developments in Las Vegas. Phil said he had some contacts in Japan who might be interested. He forwarded a very helpful contact letter with the appropriate names.

During the first week of August, Phil called to see if we had had any success seeking funds from the Japanese. So far we had received only interest, helpful suggestions, and the names of some other possible contacts for funding. Phil said he had another idea and asked us to send him the plan for the survey and the budget. He couldn't promise anything, but thought the idea was worthy of mentioning at a pitch meeting with the executive producer. Phil wanted this on his desk the first thing the following Monday morning.

The following Wednesday, Chris received a phone call from David Johnson, the executive producer. He said that the underwater survey, with divers, sonar, and video equipment, had been accepted and would be funded by *Sightings*. They would allot three days for the work and the filming, with an extra day reserved in case of bad weather.

We knew it would be a tricky balance between our quest for data and possible artifacts and *Sightings*'s desire for sensational footage, but we were elated. The dive was going to happen after all.

The next few weeks were filled with endless logistical problems. Finding an adequate survey vessel that would fit the budget and was also available for a mid-September shooting schedule was a challenge. Chris finally found a locally owned Cape Island style boat, *Murphy's Law*, hoping that the vessel's name would not prove to be prophetic. It was owned and would be piloted by Shag Harbor fisherman Bruce Addams.

Canadian Seabed Research (CSR), the company chosen to handle the survey, was having trouble guaranteeing equipment and technicians for the shooting dates. Eventually this was ironed out but other concerns cropped up, ranging from how many motel rooms would be needed to what we would do if we found something concrete. Phil Davis first put that question to us and all we could say was "I don't know." We had been so busy reading technical manuals about exotic technology like side-scan sonar, that we never paused to consider what locating an actual artifact would entail. Phil assured us that if something was found, money would be made available to deal with the find.

September 17, 1995, was the day it all came together. Chris and I, director Alec Griffith, the camera and sound crew, and the two young technicians from

Canadian Seabed Research met at the MacKenzie Motel in Shelburne. That first day provided a chance to meet informally and put together a plan so there could be some kind of balance between searching and filming. We made a special trip to Prospect Point Wharf in Shag Harbor to help the techs from CSR load their equipment and get it hooked up and on-line and generally get *Murphy's Law* ready for the sea.

One thing we remember well is how many times we retold the story of the Shag Harbor UFO crash to everyone involved in the search. We felt it was important to get everyone up to speed on the information, and received universal enthusiasm in response, from soundman to skipper. The frenzy to get everything up and ready was punctuated by visits from some of the witnesses and curious locals. News of our search had begun to leak out.

On the first day the weather was blustery, which worried us, since we had scheduled only one bad-weather day. Our skipper, Bruce Addams, assured us that the weather would improve.

Once *Murphy's Law* was loaded with the survey equipment, it was obvious that the boys from CSR were anxious to be left on their own to finish mounting and calibrating their equipment. The array of technical toys that filled *Murphy's Law*'s pilot house was impressive. Especially amazing was the navigation system that picked up information that would be integrated with our sonar data stream. A real-time differential global positioning satellite system was to be used for the survey. It operates using corrections transmitted from several reference stations in the United States. You can now get one for your automobile that will give you step-by-step directions on how to get to your destination.

We were very much concerned about target recognition in the field. The CSR techs had equipment with which they could coax out further resolution from seabed features while surveying was in progress. If possible debris or UFO wreckage appeared, we could drop anchor and dive.

Murphy's Law was also carrying a marine magnetometer, to collect magnetic readings throughout the survey. Although our little expedition did not have a remotely operated vehicle of the kind that surveyed the *Titanic* wreckage, we were equipped with an underwater video system. It was useful for quick visual checks when time or conditions did not allow time for a dive.

It was 7:00 A.M. the next morning when the phone rang in Chris's motel room. He picked it up and grunted a hello.

"Good morning, Mr. Styles, this is Don Connolly at CBC Halifax's *Information Morning*. We're live on air and I wonder if you could tell us a little of what your expedition in Shag Harbor is all about?" The cat, it seemed, was out of the bag.

So far our presence in Shag Harbor had gone unnoticed. This time things would be different. The news of our search spread across Canada. The divers and film crew, who were also staying at the same motel, had heard an interview the previous day on the same program with former *Chronicle Herald* reporter Ray

MacLeod, who had written the Shag Harbor stories of October 1967. News of our intentions had leaked out even before our arrival.

MacKenzie's Motel is a small facility of perhaps twenty rooms and six cabins. Although the motel has a swimming pool, it lacks a dining room, so meals were had elsewhere, usually at one of Shelburne Town's local restaurants. Consequently word about our presence got around fast.

The town was still in the grip of Hollywood fever following two motion pictures that had been made in the area. *The Scarlet Letter* had utilized about one third of the town, adding props and making over the old colonial-style houses along the waterfront. And if it wasn't enough to have Demi Moore in town, she was followed by Sandra Bullock, who was working in a film called *Two If By Sea* nearby. But the town wasn't too jaded to be interested in the *Sightings* episode.

The taping of our survey expedition was a little more complicated than the original segment done with director Tod Mesirow. Director Alec Griffith had a number of specific ideas about just how things would be done.

After the first day of working out bugs and laying down a few survey lines with the sonar, it was clear that Alec's concern was solely for a good take and sound bites. He did his job and he did it well, but it was obvious to us that we would have to be the ones to insure that some real survey work was done.

We decided that we had two main goals. One was

to completely survey the sound by sonar. By the sound we mean the body of water that lies between the shores of Shag Harbor and Outer Island. It was in this area where we might find evidence of the Dark Object's impact site. These were the position and the limits of the former official search.

We also wanted to be sure to check the *Navicula*'s "May not be boulders" target on the sea floor of the Shag Harbor Rip at the southern end of Outer Island.

Any remaining work time would be allocated to checking sonar hits by video drops or a dive.

While provisions were being loaded, Alec had the film crew shoot an interview with Laurie Wickens and Lawrence Smith at dockside on the Government Wharf on Prospect Point Road. Laurie, you may remember, had been the first to call in his sighting to the RCMP as a possible airplane crash, while Lawrence Smith had skippered the first boat out of this same wharf some twenty-eight years earlier.

Laurie Wickens recounted his experience on the night of October 4, 1967, and then it was Lawrence's turn to relate his role in the civilian search on that memorable night. That was followed by shots of the boat setup. Finally, by 11:00 A.M., we were ready to cast off.

We were fortunate to have Lawrence Smith along on this day. His personal knowledge and recollections were an added source of information about the sequence of events on that night in 1967. Though we always have concerns about people's memories in cases

where considerable time has elapsed, his presence
helped assure us that we were searching for something
real.

By the time we pulled away from the dock just before noon on September 18, it had been decided that
we would first investigate the mysterious *Navicula* target, which was located in an area called "the Rip"
about one and one-half miles south of Outer Island. It
was known for its fast current, due in part to the natural funneling effect of the waters pouring back out
through the sound during tidal ebb. Records showed
that was what the tide had been doing on that night
twenty-eight years earlier.

While we were gung ho for the Rip, Bruce Addams, the skipper of *Murphy's Law,* was warning us
against it. Although conditions in Shag Harbor and
into the sound were quite fair, he felt that the sea
would be much rougher there with the wind coming
out of the south the way it was. We decided to continue, and asked Bruce to give it a try. He reluctantly
swung the wheel and headed south out of the sound.

Gradually, as we came closer to the Rip, the seas
began to build up into heavy rollers and it wasn't too
long before heavy spray was being whipped by the
wind into our faces. The boat began a monotonous
rolling-and-pitching motion, while Bruce steered her
through one crest and trough after another. Once fully
into the Rip it became even worse, and it was evident
that we could not do any work there that day. We almost lost one of the Sony cameras overboard and one
of the Canadian Seabed Research techs was growing

increasingly concerned that the towfish cable might sever due to the boat's heaving and surging, and an expensive piece of equipment could be irretrievably lost.

I planted myself securely on the stern, feet braced, and was shooting the whole thing on a handheld HI-8 video camera, enjoying it all immensely. For Bruce, our skipper, and the fishermen along Nova Scotia's southwestern shore, these were typical working conditions, acceptable for fishing or slugging around lobster pots. But it was too much for scientific equipment.

We could not hope to do any work out there that day. It would be impossible to do track lines, or put a diver over the side. We could not hold position or even attempt to anchor in those seas, for fear of dragging the anchor through the very site we wished to explore, and possibly disturbing or damaging any artifacts that might be there. Reluctantly we decided to postpone this attempt until the winds and weather were more favorable.

Our second option was to return to the sound, where the waters were not nearly as rough. In the lee of Bon Portage and Outer Island, the sea was relatively calm.

Our expedition's resources were limited, especially with regard to time. This fact would cause the survey's worst disappointment. We would never get to return to the Rip. The complete sonar survey of the sound was completed on September 20, our last day on the water. Conditions beyond Bon Portage Island remained poor until the end.

While it should be noted that our only indication

that the Rip location had any possible connection to the Shag Harbor incident was the "May not be boulders" annotation on the *Navicula*'s sonar survey chart, this was still a major disappointment. Any future attempts to discover that questionable location would require considerable time laying down a sonar grid search to accurately locate and confirm the *Navicula*'s unusual target. The weather and time constraints made only one course of action reasonable, and that was to survey the sound, where the Dark Object had landed on October 4, 1967.

On that first day we were fascinated by the data as it unscrolled on thermal paper at the cost of $125 a roll, each roll lasting forty minutes. John Mercer, from Canadian Seabed Research, was busy keeping things running smoothly in the cramped quarters in the cuddy of the boat. He assured us that the data quality was good.

We saw proof of his claim when the boat passed over a submerged electrical power cable that was known to run from a fixed point on the shore over to Bon Portage Island. The cable's acoustic signature showed plainly on the sonar readouts. This was very reassuring, giving us confidence that our data was good enough to assess our targets for the video cameras and divers. However, for the rest of that day the sandy bottom was mostly featureless.

The two techs from CSR were busy, though. It was not unusual to see John spring to his feet and run to the stern of the boat to rescue the sonar towfish or magnetometer from a narrow and unexpected collision with a rapidly shallowing bottom. His navigator, Patrick Campbell, had to deal quickly with computer crashes or sudden fluctuations in GPS signals. John also had to provide additional talent as an actor, providing dialogue for sound bites and on-camera segments, while still saving the day with his technical expertise.

It was clear by the end of our first workday that days two and three would have to be longer and more productive if we were to complete our search in the sound.

I had planned for a flyover the next day, in my airplane, a Cessna 172. The plan was to remove the passenger door of the airplane and then strap Dean Brusso, the cameraman, into that seat sideways to allow filming of aerial footage. The director, Alec Griffith, would be in the back of the plane, directing and watching what the Sony camera was seeing on his own monitor.

The next morning Alec Griffith, cameraman Dean Brusso, and I drove thirty-five miles to Yarmouth Airport. I removed the right side door from the Cessna 172 and Dean was strapped into the passenger seat on that side while Alec took a seat in the back. The three of us were linked electronically by an aircraft intercom from the beginning of the flight until the end. With the prop wash and slipstream roaring past the opening at one hundred plus miles per hour, it would have been impossible for us to hear one another without the intercom.

Dean, outfitted with protective clothing to cut the cold September air aloft, seemed to take the whole thing in stride. We found out that he was always getting crammed, mounted, or strapped into one position or the other in order to accommodate the particular shooting demands made by the director.

The weather was fine for flying, like driving on a freshly paved highway. The boat spent an hour and a half trolling up and down the sound, recording more rolls of paper, while they waited for my plane, the *Kilo Julliet Julliet,* to arrive on the scene.

When the crew and I arrived overhead, we talked back and forth with the boat crew over radios, while being directed to the areas we were to film—the survey boat, the islands, and Shag Harbor. After about forty minutes we returned to the airport.

In the meantime there was more survey work to do and hopefully a couple of underwater video drops and a chance for some dives to check possible targets. A second small boat was used to get footage of *Murphy's Law* as it plied the waters of the sound. It also served as a tender between our boat and the dock in Shag Harbor, bringing or taking back people and equipment as needed.

Alec and Dean finally made it back to the boat and Alec put in a call to me while I was still at the airport. I was going to fly over the sound again a few times, solo, trying to imitate the movements the Dark Object would have made on its flight before hitting the water. I also made a visual inspection of the waters adjacent to the east and west shorelines bordering the sound. The sun's angle and the water clarity were excellent that day. I reported I could see the bottom in some areas, particularly off Outer Island, as far out as one thousand feet. The waters were shoaling very gradually, and the bottom consisted of a tan-colored gravel-and-sand mixture, dotted here and there with clumps of kelp and seaweed. Any bottom detail that aroused my interest would be checked out later. Sadly, nothing suspicious was spotted, and I turned my plane back to the airport in Yarmouth.

It was now early in the afternoon, and the boat crew once more got down to some hard work as they continued tracking sonar lines, cataloging interesting hits for a dive later in the day. Chris sat next to John Mercer, watching the data as it appeared, and, with his advice, rated targets for their "high strangeness" factor. They rated them with a one, two, or three, based on their size, symmetry, vertical height from the seafloor, and placement.

By midafternoon they had a few good targets to choose from. None were exciting but a few showed promise. The best of these was selected and *Murphy's Law* maneuvered to drop anchor as they prepared for their first dive. Alec Griffith would dive with the two professional divers who had been hired for the survey. They were equipped with underwater cameras, lights, two-way radios, and a power assist.

Once Lawrence, the dive master, was satisfied, he gave his approval and the three divers sat on the boat's rail, ready to roll over the side and into the water. Alec wanted the divers to roll over the side backwards for

dramatic effect, since the camera would be recording
it. He thought it would be even better if all three of
them rolled in at the same time.

In the first few minutes of the dive it was apparent
in the control room that there was something wrong
below. Underwater radio communications with Alec
went from confused and broken to none at all. The
other two divers lost visual contact with him. Chris
kept up attempts to reestablish radio contact with Alec
while the rest of the crew watched the surface for any
sign of his bubbles. It was starting to get scary.

After a few tense moments, to everyone's relief,
Alec's head and shoulders bobbed to the surface near
Murphy's Law's stern. Willing hands reached over the
stern, straining to get him aboard quickly. He showed no
signs of recognition at first, but after a few minutes he
got himself together and explained that a seal had failed
on his dry suit and it had filled with seawater, dragging
him to the bottom. Once Alec had gotten himself together, he had charged his buoyancy compensator with
air and bobbed to the surface. He was all right now, although he was exhausted and understandably shaken.
Fear gave way to relief aboard *Murphy's Law*.

The moment that Alec broke surface was recorded
by Dean on tape and actually made it into one of the
programs. It was shown without explanation as a filler
while the show's host explained something else.

That dive would fail to locate the sonar target. The
satellite navigation equipment had been drifting at anchor and a five-knot current was angling the divers
away from their anticipated rendezvous point. The un-

derwater camera was hung from a cable and not mounted on an ROV (remotely operated vehicle), therefore its control was limited. The camera showed huge kelp beds in the first contact area that were dense and a nuisance. Therefore the boat crew decided to move on to another target.

When they dropped anchor at the second location, the divers were ready to roll off the deck. Confidence was high that this time the sonar return represented a solid object. John Mercer conferred with Patrick and the divers, indicating that he believed the divers should be able to find something that would account for the dark image showing considerable vertical relief above the surrounding seabed.

They waited for *Murphy's Law* to stabilize in its position, with the current keeping the cable taut. Then, after a final compass check, the divers went over the side.

While the divers searched, Alec had the film crew monitoring the boat's communications with them. Our underwater camera allowed an occasional glimpse of their actions. The target on this dive would prove to be an unusual rock formation. Once the divers were aboard, the boat crew went back to laying down more lines of sonar tracking in the sound. It was apparent at this point that the sound would be completely surveyed before the third working day came to an end. But day two was yet to end and it had been a long and stressful one. They took a breather and had something to eat.

By the time darkness approached, they had scanned eighty percent of Shag Harbor. They had discovered a few interesting targets, but nothing that jumped off the page, and nothing to indicate anything of extraterrestrial origin.

A night dive was planned to give *Sightings* some dramatic footage. A target was chosen that showed sharp lines on the leading edge of the scan and a hint of symmetry. It stood a chance of being something other than a natural formation.

It was a beautiful evening with perfect visibility. The night sky was awash with stars that city dwellers don't usually get to see. But nothing unnatural was to be found under the water.

It was close to ten that evening when the boat headed back to shore. When *Murphy's Law* steered in behind the wharf at Prospect Point, a black pickup truck started its engine and sped down the long jetty and away. The skipper, Bruce Addams, mentioned that the same truck and driver had been there earlier that day as we were loading supplies aboard. He could not identify the man, nor was he known to any of the others.

Apparently a rumor was beginning to spread through the small communities around Shag Harbor that we had found something. By the time we arrived back in Shelburne, the rumor had proceeded us. The town was buzzing that we had found something for sure. Why else would we be out so late at night on the water, sneaking around in the dark? Even Laurie Wickens was suspicious, and it was obvious that denial alone would not quench the fires of speculation. For once I found myself in the position of those officials who grope for a plausible denial.

We had a discussion with Alec Griffith after supper about a UFO sighting that Lawrence Smith, one of the Shag Harbor witnesses, had had in 1970. He agreed to talk and described the strange-looking foam that had been brought up from below.

Chris had found a police report about a UFO sighting that Smith had with his brother Gene in 1970. They noticed a small orange light hovering in the air just over the trees. Soon five more lights appeared, together forming a pattern somewhat like the Big Dipper, with the "handle" missing.

As one of the lights winked off, their car engine died and their tape deck quit playing. Later, it seemed to the brothers as if they had lost about ten minutes of time.

They made a report to RCMP Constable Ralph Keeping, who was a friend of Lawrence's. We were later able to locate and read this report to Lawrence. One part of it was new to him: Keeping hadn't told him that he had passed on the information to Colonel Rushton of NORAD.

Alec was impressed by this unusually solid sighting at Bear Point. Our discussion would later be interpreted by Alec as a suggestion for the show, and would lead, by the end of the third day, to a classic breakdown in communications.

September 29 was to be our third and final workday on the water. I was aboard again, having missed the previous day while in the air. In only a few hours we finished scanning the sound. No targets of a highly unusual nature were discerned during our interpretation of the scrolls. It was decided that the afternoon would be spent either making another attempt at locating the *Navicula* target on the Rip, or doing one more dive for the purpose of shooting additional footage for *Sightings*.

After a lunch break Bruce Addams, the skipper of *Murphy's Law,* informed us that the weather was going to deteriorate and that the Rip would not be the place to be when it did. It was obvious that the weather was going down. Using a cell phone, I checked with Flight Services in Yarmouth, because I, too, was worried about the weather.

After some speculation I decided I would take the dory back to shore and, with Lawrence Smith's help, get to Yarmouth, pick up my plane, and fly it back to Halifax. Otherwise the aircraft would be pinned down in Yarmouth for several more days. You have to worry about airplanes almost as much as you worry about your children.

Meanwhile, the technical crew from CSR got back to work, while the divers suited up, preparing for their final dive, an exercise primarily to obtain underwater film footage. However, it was agreed that it would be done in an area where several minor targets had been detected. Lawrence, our dive master, planned to cover as much area as his air supply would allow. Eventually the divers surfaced several hundred feet away from our boat with nothing unusual to report.

Before the dive ended, however, we had a visitor. A Cessna flew in from the west and over the boat, circling us in a steep bank. Someone was filming us with a large professional camera. This continued for about five minutes before the pilot waved and left the area. We decided the plane had probably been chartered by one of the local television affiliates.

When the dive ended and all three of the men were back aboard, the weather deteriorated as if someone had thrown a big "nasty" switch. *Murphy's Law* plowed through the chop gamely as we made our way to shore. We were all looking forward to a hot supper, as well as some time to reflect. Supper was not long in coming; however, reflection would have to wait. Interviews, taping, and more problems lay ahead before this little expedition could be considered officially over.

During the evening meal director Alec Griffith told us that day four would be spent interviewing Lawrence Smith and producing a segment about the Smith Encounter near Bear Point Road.

We were stunned. Alec had called Hollywood and pitched the idea as worthwhile use for the scheduled "bad-weather day." The supper exchange became what might be called a communications breakdown. Statements were made such as "Don't jeopardize this important investigation!" And of course the always unfortunate "Don't you f— with me."

We were also concerned because Lawrence had told us that he did not want to go public with his story. We believe that one reason we've gained some respect from eyewitnesses in investigations is that we respect a person's limits. No means no, and we accept this.

By day four cooler heads prevailed. We had a good talk with Alec, and Lawrence did an interview. The expedition packed up and left Shelburne for the drive back to Halifax and Canadian Seabed Research's offices in Porter's Lake on Nova Scotia's eastern shore. Here Alec did two lengthy interviews about Shag Harbor and what had been accomplished to date.

We all said our good-byes in CSR's driveway and went our separate ways.

What did it prove? Some would suggest that our September 1995 search proved that nothing significant had happened. But as we see it, our expedition added weight to the evidence that the Dark Object somehow ended up elsewhere. This fact says much about its mysterious origin and capabilities. If it were a crashed airliner, it couldn't have moved away underwater.

As far as we're concerned, the search is not yet over.

THE MYSTERY CONTINUES

On the sixteenth of March, 1997, we made a phone call to a man we hoped was the person we had been trying to locate for three years. It was the second call we had made in as many days. The first had not panned out. The name was the same, but he was not the man we were looking for. The second time, however, we hit pay dirt. I located him through a program on the Internet.

Barry Crowell was the lighthouse keeper who we believed was on duty at the Cape Roseway lighthouse during the events that transpired off Government Point at the same time as the Shag Harbor incident. Since the lighthouse was located only two-and-a-half miles across the waters of the Eastern Way of Shelburne

Harbor from the sub-tracking base at Government Point, it enjoyed an excellent vantage point from which to view the area in question.

A woman answered the phone and I asked if I could speak to Barry Crowell. She called to someone and seconds later a man answered. That special instinct that we all seem to have kicked in and we knew without asking that this was going to be our man.

After introducing ourselves and exchanging a few preliminary remarks, we began to ask the man some questions, the most important of which was whether he had been on duty the night of October 4, 1967. Barry answered hesitantly at first, as most people do when someone asks questions of them about their past, particularly if that someone is a total stranger. He said he could not be sure of exact dates though he thought he might have been on duty that night.

We began by asking if he remembered a flotilla of ships, six or seven, being anchored offshore during this period, in sight of both Government Point and Cape Roseway. No, he did not remember anything like that, although that might have been due to heavy fog that was frequently present. When asked if he would remember foghorns blowing on the ships if there was fog, he replied that he would probably have taken no notice of them. We pressed him further, asking if anything out of the ordinary had happened, maybe something he considered not related.

"Well"—he hesitated—"there was that night when the commandos came ashore and took over the lighthouse."

"Commandos?" This caught us completely off-guard. "What do you mean, they took over the light-house?" Barry really had our attention now. The story as he told it came out in little snippets, as he recalled events that had unfolded over a period of about three days.

After we had most of the details, we were able to piece together what had happened. This is what he recalled.

It was about ten minutes to midnight on the night in question. Barry Crowell, as the junior lighthouse keeper at Cape Roseway, was pulling the graveyard shift that evening. He was on his way from his living accommodations to the lighthouse on McNutt Island. In a few moments he would relieve Brenton Reynolds, the light keeper on the backshift. He in turn would be relieved at 8:00 A.M. by Harry Van Buskirk, the third keeper of the light.

The night was cool and foggy, with clear patches. He could hear breakers crashing against the rocky shore periodically, indicating a heavy swell was running that night, no doubt being influenced by the tides and the currents.

When he was only a hundred feet or so from the lighthouse, Barry met Brenton on his way home. They stood and talked for a moment. Quite suddenly and without warning lights exploded above them, one, two, three of them, yellow-orange in color and ringed in haloes from the fog. Near darkness was turned to daylight.

At first he was stunned, perplexed, and a little fearful until he recognized them as flares drifting

downward on parachutes, sputtering and dripping
residue as they neared the ground. The whole area was
drenched in an eerie orange mantle of flickering light
and shadow. These were followed by several more.

Barry's next impulse was to lower his gaze and
seek out the source of the flares, his thoughts turning
to a possible craft or vessel in distress. He scanned the
nearby sea and the rocky shoreline and was presently
rewarded with the scene of a rubber Zodiac boat being
paddled ashore and in some distress.

Barry remembers heading for the shore with Brenton and thinking that whoever these boatmen were,
they had picked the worst possible piece of coast on
which to make a landfall. He was in time to see the Zodiac spill most of its human cargo into the heavy seas
when its bow came in contact with the rocky shore.

Barry and his friend ran to the shore to offer assistance and discovered some very frightened men
dressed in strange, dark military garb. His first thought
was that they might have been Russians. This was not
impossible, since Russian fishing trawlers were in evidence off the Nova Scotia coast in fairly frequent numbers in the 1960s. In those days there was only a
twelve-mile limit around the Canadian coast. No one
really believed that these vessels were there to fish,
though, since they were usually festooned with anten-
nas and radar equipment that would have been the
envy of local fishing boats plying those waters.

Barry was only slightly relieved to hear some of
the five or six men they helped ashore speaking En-
glish, even though it was a very British style of En-

glish. The men were soaked through and their leader explained that this was a mock commando raid and that they were there to secure the light station and the operators. As an exercise they were to hold the island and control the light station operation and its radio traffic.

Barry and his wife Donna recall this crew as having been there for the best part of two days, if not three. They were supplied from the air by helicopters, which Barry and Donna heard but did not see, farther down the island. Though these "commandos" received rations, they were more inclined to eat the food offered by the Crowells, and for a couple of days some even wore pants borrowed from Barry while their own clothes dried out. The strangers asked the Crowells not to tell any other military personnel about this.

Barry remembers a Piper-Cub-type airplane circling over the area constantly for several days, as well as regular helicopter visits. When I asked how this could be so with so much fog present, Barry explained that it was a peculiarity of the area that fog would build up on the surface of the water to the height of fifty to seventy-five feet, blocking horizontal but not vertical visibility, affecting boats but not planes.

We are trying to track down the light station records from that period, which will at least give us some indication as to whether there was fog on the night of October 4 in that area. Our experience indicates that these records probably would not have been kept for thirty years.

Barry asked the "commando" leader on one occasion about where their Zodiac had been launched from and was told they had come from a submarine, but he was unable to remember whether he was given a name. When I interviewed him in person, I asked him if he thought these guys were really commandos. He laughed and shook his head and said, "No way. If they were, they were poorly trained. These guys were really frightened when they came ashore. They thought they were going to drown."

I said that I thought it peculiar that commandos would fire off flares at night when they were trying to sneak ashore undetected.

Barry agreed. "These guys did everything but radio ahead. They certainly weren't quiet about it."

We asked Barry what he thought this whole adventure had been about. He suggested that it might have been part of some NATO exercise. He may be right.

One notable detail was later brought to light by Chris. Earlier in 1967 the Canadian government had decided to get into the submarine game and purchased three used "Ranger Class" subs from the United Kingdom. The first one, the *Onandogan*, was brought over to Canada, bound for the Royal Canadian Naval Base in Halifax, Nova Scotia, and due to enter port on October 4 to 5. While still several hundred miles out to sea it was diverted to Shelburne Harbor. Back then only two or three of its crew were Canadian and knew anything about submarines. For the most part the officers and crew were British.

But a NATO exercise is not the only possible explanation for what happened that night. Barry Crowell's wife was sure she remembered hearing about the

Shag Harbor UFO crash on the radio about the same time the "commandos" came ashore.

This made us wonder if the raid could have had something to do with the naval search for the Dark Object. Could these soldiers have been looking for survivors of the crash of a UFO—perhaps nonhuman survivors? We don't know the answer to this question, but more information may turn up later, as we continue to turn over more stones in our ongoing search for documentation about the Dark Object.

Another interesting twist to this case has come to our attention recently. A week after the "commando raid," local fisherman Lockland Cameron observed a strange object barely a half mile away from where the Dark Object was reported to have entered the water.

On Wednesday evening, October 11, 1967, Lockland, his wife Lorraine, and their daughter Louella were watching television in the living room of their house in Woods Harbor, Nova Scotia. That night they were entertaining Lockland's brother Havelock, Havelock's wife Brenda, and their child.

Suddenly the TV picture began to wiggle, with a white line down the middle of the screen. Lockland endured this for a few moments, then went to the window to look for an airplane, which in those days was the usual culprit when your TV picture was shaky.

The Cameron house had a straight view down the sound. Cameron's attention was drawn to a group of brilliant red-orange lights in the sky, close to the water. He called to the others, and they all hurried outside to get a better view.

The six family members stood and watched, fascinated by the lights. First there were six of them, then four, flashing in sequence to the left, then back to the right in reverse order. The object itself was stationary, hovering at an altitude of about five hundred to six hundred feet.

They watched the lights for seven to eight minutes, until they suddenly went out. Shortly afterward they reappeared in a slightly different area, but this time they pointed downward at an angle of about thirty-five degrees. The family continued to watch the object until it vanished over Tusket Island. Moments later they saw the lights again, only this time there appeared to be two objects, speeding rapidly over the Gulf of Maine toward the United States.

Lockland went inside and called the RCMP. Because the first lights he'd seen were so close to the surface of the water, he thought that the first object they'd seen might have come up out of the water, and he thought the Mounties might be interested.

A Mountie came to the house and took the statements of the Cameron family. Later he called the Royal Canadian Air Force and found out that there were no reported aircraft in the area and no known air operations in progress. The Mountie's report read, "Suggest government personnel interview persons concerned above and those in sighting on 4 Oct. 67."

Only seven days earlier, at about the same time of night, a UFO of approximately the same size and with the same lighting pattern had supposedly crashed into the waters only a mile or so to the south. Could the

Dark Object have rested on the bottom of the sound for a week, then left again, the way it had come? This would explain why no object was ever found during our diving expedition.

While writing this book, I received an e-mail from a Ms. Fountain in Montreal, Canada. She had read a magazine article I wrote on the Shag Harbor incident, and wanted to relate a story that had been told to her by her father in 1967, when she was ten years old. They lived in Pubnico at the time, which was another fishing village about twenty miles northwest of Shag Harbor.

Her father, Wayne Nickerson, had just been reading the *Chronicle Herald* story about the incident in Shag Harbor, which had taken place three nights before. Mr. Nickerson told her about an experience he had had on the night of October 4 while returning from his job as a railroad switchman. He was traveling west from Shag Harbor to Woods Harbor, when he saw something strange in the sky. He pulled his car off onto the shoulder of the road, cut the engine, and got out to observe more carefully. He was certain that his eyes were playing tricks on him.

In the sky he saw two moons. He stared at them, as they stayed fixed in place for a few seconds, then one moon dropped swiftly downward and landed gently on the sound, where it drifted silently.

Greatly disturbed, Nickerson got back in his car and left the area in a hurry. He mentioned the incident to no one until Saturday the seventh, when he read the article in the *Herald*.

Ms. Fountain said this type of story was totally out of character for her father, who was a plain speaker, not given to telling stories or adding embellishment. He died in 1991, but about six months before his death he recalled the incident once more. He hadn't spoken of it in twenty-four years.

What especially interested us about this story was that there was no moon at all in the sky on October 4, 1967, the night of the Shag Harbor incident. There was only a new moon that set before sundown that night—a fact that we've checked out carefully. So Mr. Nickerson not only saw one moon that shouldn't have been there—neither moon he saw should have been visible in the sky. This brought up the idea that there could have been two UFOs that night.

Perhaps the second object he saw in the sky, after the first one dived into the sound, was the object seen by Norm Smith and his father, many minutes after Laurie Wickens saw an object go behind the trees near the Irish Moss plant. This object could have crashed, or dived, into the sea at Shelburne.

There is some evidence to suggest that, in fact, two UFOs did go into the sound that night. Earlier on in the investigation we noted that Norm Smith saw the lights of the UFO descending toward Shag Harbor the second time. After Dave Kendricks dropped him off, he quickly ran into the house to get his father so he could point them out to him. They both agreed that the object was in danger of crashing somewhere near the fishing village.

They climbed into their car to go investigate, and

while they were backing out of their driveway, they had to apply the brakes to avoid crashing into Corporal Werbicki's police cruiser, with its flashing red lights, which was already headed out to the Irish Moss plant, where Laurie Wickens was keeping an eye on the object he had seen out in the sound.

How could this be? The only explanation is that there were actually two Dark Objects. The Cameron family also saw two objects racing away in the sky, a week after the crash.

Recently another interesting piece of news came our way. Acting on a tip UFO investigator Steve MacLean gave to Chris Styles, we made a call to a military officer we'll call Leo.

The phone rang on the other end, four or five times. We were just about to hang up when a woman answered. We asked to speak to Leo. He came to the phone after a few seconds. We identified ourselves and explained that a fellow UFO investigator had passed his name on to us as a possible information source for the Shag Harbor incident.

"Oh, yes," he said.

Not wishing to lead him, we asked if he'd had any involvement with the investigation that night. We already knew some of it, but were fishing for more.

"I was one of the Royal Canadian Air Force radio officers that night. I was handling the military traffic. There was much excitement over this event."

We knew this much but wanted to see if he had any more details of the events that night that he could share with us.

"You realize that I can only go so far?" he asked.

"I know you have the National Security Act to be concerned with. You probably signed an oath or something."

"Oh, I'm not worried about that. It's the code that I have to adhere to."

"Code?"

"Yeah, the code that we in the armed forces abide by. If they say don't talk beyond a certain point of information, then we don't."

I was tempted to ask about the code concerning the civilian population's right to information, since we're the guys who pay the bills, but we didn't want to get into an argument. Instead we asked, "What can you tell me about the night of October 4, 1967? The fact that an object crashed into Shag Harbor is pretty well documented."

"Oh, yes. But you know there were two objects that went into the water that night, not just one. The divers took all kinds of pictures."

This new theory is breathtaking, and could explain why both Shag Harbor and Shelburne were involved. We could think of hundreds of questions to ask him, but he wouldn't tell us anything more.

And so, the mystery continues.

ROYAL CANADIAN MOUNTED POLICE - GENDARMERIE ROYALE DU CANADA

OTHER FILE REFERENCES: REF. AUTRES DOSSIERS:	DIVISION "H"	DATE 7 OCT 67	RCMP FILE REFERENCES: REF. DOSSIERS GRC:
	SUB DIVISION - SOUS-DIVISION HALIFAX		67-400-23-X
	DETACHMENT-DÉTACHEMENT LUNENBURG		

RE:
OBJET:

Unidentified Flying Object.
Sighting of -
Sambro Light, N.S.
(4 OCT 67)

1. On this date a request was received from the Halifax Sub-Division Section N.C.O., via XJD 84, to contact Capt. Leo Howard MERSEY, of the M/V "J.B. NICKERSON", relative to his sighting of a flying object off Sambro Light on 4 OCT 67. It was further requested that the results of enquiries be forwarded to Barrington Passage Det., in view of a similar sighting in that area.

2. Capt. MERSEY was interviewed and the following statement obtained:

STATEMENT OF CAPT. LEO HOWARD MERSEY (B: 12 JUNE 22),   Centre, N.S.
Centre, Lun. Co., N.S.   7 OCT 67.

At about 9 P.M., on the 4 OCT 67, I noticed an object with three flashing red lights. Radar indicated this object to be sixteen (16) miles away. It was very clear that night and we could see the lights of Halifax. At the time our boat was 32 miles south of the Sambro Light and the object was approximately 16 miles north east of us. I would say the object was 16 miles south east of Sambro. At times the Navy do a lot of practising in the area. At the same time there were three other objects on the radar and about 6 miles from the first object. I would say it disappeared about 11:00 P.M., when it went up in the air. I could not see any shape or form to it because of the distance. When it went into the air it only had one flashing light. While the object was on the water, or close to the water, it had three real bright flashing red lights. All the lights on it were red. I could not see any lights on the other three objects as they were only appearing on the radar. It is not unusual to see the Navy, or aircraft, dropping things into the water there. I had never seen anything like that before but it sounds like the thing they are looking for down off Shelburne or Barrington Passage. When the object left it went straight up in the air with only one red light.

Witnessed: D.J. RAHN, 2/Cst.   Signed: Capt. Leo H. MERSEY.

3. Capt. MERSEY is considered to be a reliable type individual and bears a good reputation in his community.

4. Barrington Passage Detachment were advised of the foregoing via telephone. A copy of this report is being forwarded direct to that point for their information.

CONCLUDED HERE.

Cpl.
(J.F. Kovacs) #18905.
I/C Lunenburg Detachment.

C O P Y

PRIORITY

FM: BARRINGTON PASSAGE DET.

TO: HALIFAX S/DIV

16 UNIDENTIFIED LIGHT REPORTED SIGHTED 8:30PM OVER SOLOMONS
ISLAND ONE MILE WEST OF LOWER WOOD HARBOUR SIGHTED BY LOCKLAND
MACLEAN CAMERON AGE 41. LORRAINE GERTRUDE CAMERON 34. HAVELOCK
MCGREGOR CAMERON 33. BRENDA MARY CAMERON 28 AND TWO CHILDREN.
LIGHT IN STRAIGHT LINE IN SOUTHEAST DIRECTION AND 45 DEGREE ANGLE.
NO FORM SEEN, ESTIMATED SPREAD OF LIGHT TO BE 50 TO 60 FT. ON
FOUR OCCASIONS THE LIGHTS WHICH WERE BRIGHT RED AND AS BRIGHT
AS LOCAL BUOY LIGHTS WENT OFF IN SEQUENCE FROM BACK TO FRONT AND
THEN WERE LIT IN SEQUENCE FROM FRONT TO BACK. THE FIRST TIME
THERE WERE SIX LIGHTS AND LAST TIME FOUR. LIGHTS HOVERING
AND SHIFTING FROM SIDE TO SIDE. RED LIGHTS WATCHED FOR 7-8
MINUTES AND WERE EXTINGUISHED AND YELLOW-ORANGE LIGHT APPEARED
AND DISAPPEARED OVER THE HORIZON TO THE NORTHWEST. THIS LIGHT
OBSERVED FOR APPROX 10 MINUTES. ALL PERSONS SOBER AND APPEAR
TO BE SINCERE. SIGHTING ALSO REPORTED TO RCAF BACCARO. UNDER-
STAND BACCARO RADAR NEGATIVE. NO KNOWN OPERATIONS IN AREA.
SUGGEST GOVERNMENT PERSONNEL INTERVIEW PERSONS CONCERNED ABOVE
AND THOSE IN SIGHTING ON 4 OCT 67.

ELS.

ROYAL CANADIAN MOUNTED POLICE - GENDARMERIE ROYALE DU CANADA

9 DEC 70

Halifax

70-100-15

Barrington Passage

Unidentified Flying Object, Bear Point, Shel. Co., N.S. 25 NOV 70.

25 NOV 70

1. At approx. 9:50 pm this date while on patrol at Barrington Passage, N.S. I was contacted by Lawrence Charles SMITH of Lower Shag Harbour, Shel. Co., N.S. SMITH was accompanied by his brother Manus Eugene SMITH of the same address and they stated that they had seen lights in the sky at Bear Point, Shel. Co., N.S.

2. They stated that they had seen five or six lights in the sky approximately 30 to 50 feet in the air. These lights were described as being 20 inches in diameter and were reddish orange in color. These two witnesses stated that these lights disappeared without moving. The duration of the sighting as described by these witnesses was approx. 20 to 30 seconds. They stated that the tape player in the car had stopped and that the car had stopped on its own when it had been travelling approx. 45 M.P.H. They stated that the second light from the left had gone out and then lit up again at least once.

3. An immediate patrol was made to the scene with the SMITH brothers. They explained the position of these lights and how they had been driving normally along the highway towards their homes when the sighting occured. Nothing could be seen at the place of the sighting out of the ordinary; the area is thickly wooded. I was the scene for approx. 30 minutes and after a thorough search of the area I left.

26 NOV 70

4. At approx. 9:15 am this date Col. RUSHTON, C.O. of C.F.S. BARRINGTON at Baccaro, Shel. Co., N.S., called and asked if our office had any reports of U.F.O. sightings the previous night. He was advised that there had been. During a short telephone conversation he revealed that one of his men had seen a U.F.O. the previous evening and from comparing notes it appeared that the sighting had been very similar to the sighting reported by the SMITH brothers. Col. RUSHTON stated that he wished to see me after I had conducted my investigation concerning the U.F.O.

5. Lawrence Charles SMITH and Manus Eugene SMITH of Lower Shag Harbour, Shel. Co., N.S. interviewed separately concerning the sighting. They related to me how they had seen the lights as described in paragraph No. 2. statements attached. Diagram of Lawrence Charles SMITH'S version of the sighting attached.

6. Patrol made to C.F.S. BARRINGTON and the sighting was discussed with Col. RUSHTON. He was supplied with the names of the witnesses already interviewed.

CONTINUED ON PAGE #2.

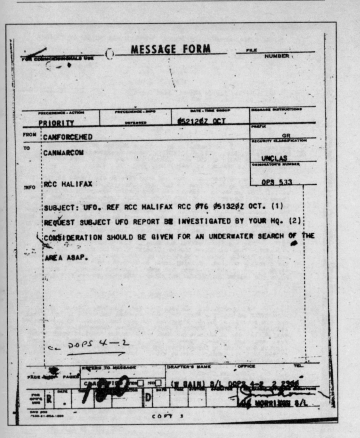

MESSAGE FORM

FOR COMMUNICATIONS USE

FILE
NUMBER

PRECEDENCE - ACTION	PRECEDENCE - INFO	DATE - TIME GROUP	MESSAGE INSTRUCTIONS
PRIORITY	DEFERRED	Ø521207 OCT	

PREFIX

FROM CANFORCEHED

GR

SECURITY CLASSIFICATION

TO CANMARCOM

UNCLAS

ORIGINATOR'S NUMBER

INFO RCC HALIFAX

OPS 533

SUBJECT: UFO. REF RCC HALIFAX RCC Ø76 Ø51320Z OCT. (1)
REQUEST SUBJECT UFO REPORT BE INVESTIGATED BY YOUR HQ. (2)
CONSIDERATION SHOULD BE GIVEN FOR AN UNDERWATER SEARCH OF THE
AREA ASAP.

DOPS 4—2

(W BAIN) S/L DOPS 4-2 2 2306

MORRISON S/L

COPY 2

Whitley Strieber is the author of *Communion, Confirmation,* and the novel *The Hunger*. His most recent book, *The Coming Global Superstorm,* was a national best-seller. He hosts the Sunday night radio program *Dreamland*. Mr. Strieber was born and raised in Texas and now lives in San Antonio.

Don Ledger, author and private pilot, works as a director for Legislative Television Broadcast Services for the Province of Nova Scotia. He is the author of two other books, *The Maritime UFO Files,* concerning UFOs in Eastern Canada, and *Swissair Down,* which deals with the crash of Flight 111 off Nova Scotia in 1998. Don and his wife, Gail, live in Bedford, Nova Scotia.

Chris Styles is a UFO researcher living in Dartmouth, Nova Scotia. Chris has been researching UFO events since the early nineties.